REVISE AQA GCSE
Geography A
For the linear specification first teaching 2012

D1146474

REVISION GUIDE

Series Consultant: Harry Smith

Author: Rob Bircher

Also available to support your revision:

Revise GCSE Study Skills Guide   9781447967071

The **Revise GCSE Study Skills Guide is** full of tried-and-trusted hints and tips for how to learn more effectively. It gives you techniques to help you achieve your best – throughout your GCSE studies and beyond!

Revise GCSE Revision Planner   9781447967828

The **Revise GCSE Revision Planner** helps you to plan and organise your time, step-by-step, throughout your GCSE revision. Use this book and wall chart to mastermind your revision.

> For the full range of Pearson revision titles across GCSE, BTEC and AS Level visit:
>
> www.pearsonschools.co.uk/revise

ALWAYS LEARNING

PEARSON

Contents

A small bit of small print

AQA publishes Sample Assessment Material and the Specification on its website. This is the official content and this book should be used in conjunction with it. The questions in *Now try this* have been written to help you practise every topic in this book. Remember: the real exam questions may not look like this.

Unstable crust

At the centre of the Earth is the **core**. Surrounding this is a layer of **molten** rock called the **mantle**, then on the surface is a thin layer called the **crust**.

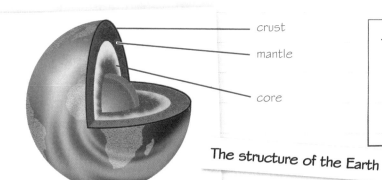

— crust
— mantle
— core

The structure of the Earth

There are **two** types of crust.
- **Oceanic crust:** denser and thinner (5 km thick).
- **Continental crust:** less dense, but thicker (30 km).

Tectonic plates

The Earth's crust is made up of seven large tectonic plates and lots of smaller ones.

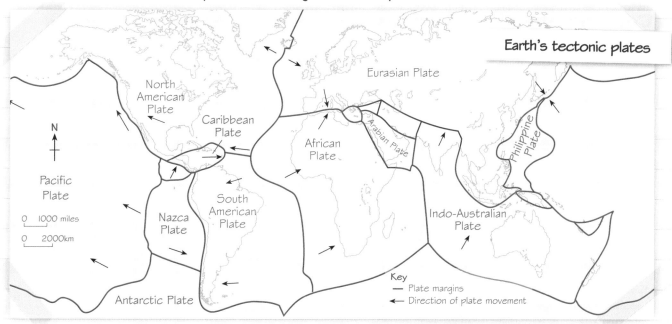

Earth's tectonic plates

North American Plate

Eurasian Plate

Caribbean Plate

African Plate

Arabian Plate

Philippine Plate

N

Pacific Plate

0   1000 miles

0   2000 km

Nazca Plate

South American Plate

Indo-Australian Plate

Antarctic Plate

Key
— Plate margins
← Direction of plate movement

Moving plates

The crust is unstable because the plates may move apart or move closer together or slide past each other. Some of these movements lead to the formation of, for example, **fold mountains** and **volcanoes**.

Worked example

Explain how tectonic plates move. **(2 marks)**

The heat from the Earth's core creates convection currents in the mantle. This hot molten rock rises moves the plates on the surface which form the crust.

Questions that ask you to 'Name' something are not looking for description or explanation.

1 Name **three** tectonic plates. **(3 marks)**
2 Iceland sits on the boundary of which two plates? **(1 mark)**

Plate margins

There are three types of plate margin: destructive, constructive and conservative.

Destructive plate margins

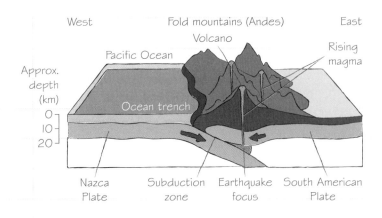

A destructive margin is when plates move together. The lighter continental crust stays on top while the denser oceanic crust is pushed down into the mantle where it melts.

Features:

- Rising magma can cause **composite volcanoes**.
- Energy builds in the subduction zone and is released as **earthquakes**.
- **Fold mountains** are formed due to the impact of the colliding plates.
- **Ocean trenches** are formed.

Constructive plate margins

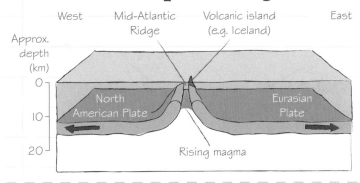

A constructive margin is when plates move apart and magma rises to fill the gap.

Features:

- Usually found under the sea.
- Ridges and **shield volcanoes** are formed by the build up of **magma**.
- **Volcanoes** can form islands if they rise above sea level.

Conservative plate margins

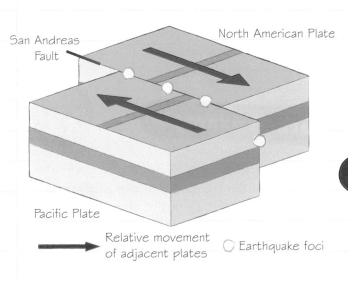

A conservative margin is when plates slide past each other. Sometimes the plates are going in the same direction but at different speeds. At times they stick together.

Feature:

- Earthquakes occur when built up pressure forces plates unstuck.

Worked example

tier **F**n

Iceland is a volcanic island located on the mid-point of the Atlantic Ocean. What sort of plate margin is Iceland sitting on? **(1 mark)**

Iceland is sitting on a constructive plate margin.

Now try this

1 Name one example of a conservative plate margin. **(1 mark)**

2 Give two ways in which earthquakes are triggered by plate movements. **(2 marks)**

tier **F&H**

Fold mountains and ocean trenches

Fold mountains and ocean trenches are landforms that occur at destructive plate margins.

Fold mountains

The formation of fold mountains

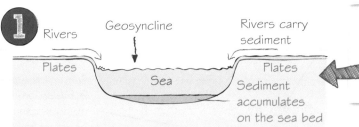

First, thick sediment layers form in huge depressions called geosynclines under the sea.

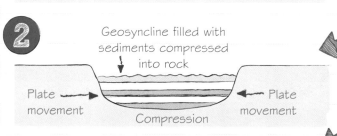

Over millions of years the sediments are compressed into rock.

The sedimentary rocks are forced into folds as the plates move.

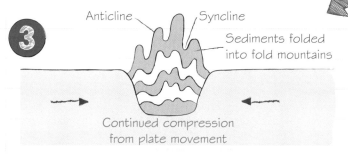

Fold mountains have been created at different times in the Earth's history. Mountains such as the Alps, Himalayas and Andes are called **young fold mountains** because they were formed in a recent mountain building period.

Ocean trenches

These are formed in the subduction zone of a destructive plate margin. One wall of the trench is formed by the subducting oceanic crust. The other wall is the edge of the continental crust.

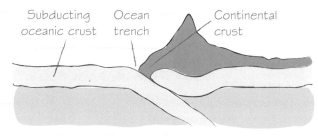

Worked example

tier **H**

Why are ocean trenches of little use to humans?
(2 marks)

Ocean trenches are extremely deep: too deep for humans to be able to reach the ocean floor (e.g. to drill for oil), except for extremely short times using extremely expensive machinery.

For a question like this you can make two brief points to explain your answer

Now try this

Explain the difference between a syncline and an anticline.　　**(2 marks)**

tier **F&H**

Fold mountains

The Alps are an example of a fold mountain.

Traditional farming moved livestock up to the high pastures each summer while hay was grown on the rich soils of the valley floors. Now farmers buy in fodder crops and pasture livestock on the valley bottoms all year.

Summer tourism is based on the dramatic mountain scenery.

Case study

You need to know a case study of one range of fold mountains – how they are used and how people adapt to the conditions.

Travel between valleys used to be long and difficult. Now railway tunnels have been cut through the mountains to make travel much quicker and easier.

Winter tourism is based on skiing because of the steep, snowy slopes.

The steep valley sides and fast-flowing mountain rivers are perfect for HEP.

Fields cultivated for crops in summer

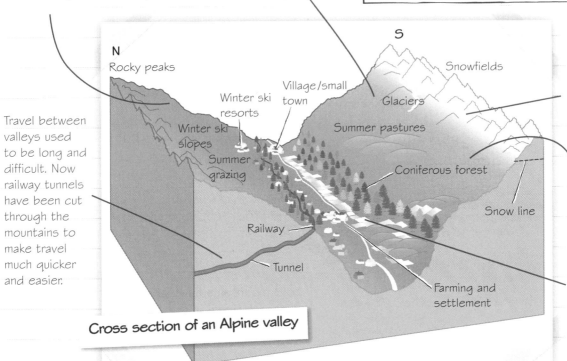

Cross section of an Alpine valley

Labels: N, Rocky peaks, Winter ski resorts, Winter ski slopes, Summer grazing, Village/small town, Railway, Tunnel, S, Snowfields, Glaciers, Summer pastures, Coniferous forest, Snow line, Farming and settlement

Challenges of living in the Alps

- **Relief** – high mountains, steep slopes, narrow valleys with little flat land for farming.
- **Climate** – temperature drops with height. Growing season is short.
- **Soils** – stony, thin and infertile.
- **Accessibility** – travel along valleys is OK but between them is very difficult!

Worked example

tier **F&H**

Give **one** example of how people in a fold mountain region have adapted to the problems of steep relief and poor soils. **(2 marks)**

In the Andes, traditional farming uses terraces to create areas of flat land on steep slopes. The terraces retain water better and the soils can be deeper and more fertile.

Use the template on page 116 to help you make an index card to revise your case study. You could use these headings: Challenges; Uses; How Humans adapt.

EXAM ALERT!

Some students missed that this question asked about both steep relief **and** poor soils.

Students have struggled with exam questions similar to this – **be prepared!**

Use the template on page 116 to help you make an index card to revise your case study.

Now try this

Why is HEP often associated with fold mountain regions? **(2 marks)**

tier **H**

Volcanoes

There are two main types of volcano, and they have different characteristics.

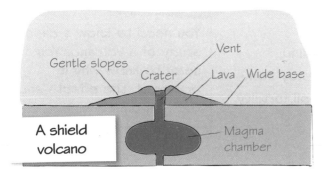

A shield volcano

A composite cone volcano

Location: constructive plate margins.

Formation: magma rises to fill gap between plates.

- not explosive
- regular and frequent eruptions
- runny lava
- pours out smoothly
- creates gentle slopes and wide base

Characteristics

- made of lava only

Example: Surtsey island, near Iceland.

Location: destructive plate margins.

Formation: magma in the subduction zone is under great pressure and may be forced up to the surface through cracks.

- subsidiary cones
- made of lava and ash

Characteristics

- irregular eruptions with long dormant periods
- viscous lava
- violent eruptions
- tall cone, steep sides, narrow base

Example: Etna in Italy, Krakatoa in Indonesia.

Supervolcanoes

A supervolcano eruption is 1000 times bigger than a normal volcano. So much dust is ejected into the atmosphere that global cooling occurs. After a supervolcano erupts, the magma chamber collapses, forming a caldera. The last known supervolcano was 75000 years ago.

Worked example

tier H

Describe **two** differences between shield volcano eruptions and composite volcano eruptions.

(4 marks)

Shield volcano eruptions are regular, frequent and not violent, whereas composite volcano eruptions are violent and irregular. Shield volcano eruptions produce a low, shield-shaped cone with a wide base. Composite volcano eruptions produce a steep-sided, tall cone with a narrow base.

EXAM ALERT!

Avoid direct opposites: e.g. 'a composite volcanic eruption is violent, a shield volcanic eruption is not'.

Students have struggled with exam questions similar to this – **be prepared!**

Now try this

tier F n

tier H

1 Give an example of a shield volcano and a composite cone volcano. **(2 marks)**
2 Describe the characteristics of a supervolcano. **(2 marks)**

Volcanoes as hazards

When a volcano erupts it can have both positive and negative impacts.

Primary effects:
- people injured or killed
- buildings and farmland destroyed
- communications disrupted.

Secondary effects:
- costs of rebuilding
- tourists might stay away
- ash improves soil fertility.

Positive impacts:
- 👍 fertile soils
- 👍 tourist attractions
- 👍 natural hot springs.

Negative impacts
- 👎 always destructive, even shield volcanoes
- 👎 global disruption, e.g. ash clouds disrupt travel
- 👎 unpredictable so it is difficult to prepare for them.

Case study

Soufriere Hills volcano, Montserrat, 1995 and 1997

Location: The island of Montserrat in the Caribbean which is on a destructive plate margin

Cause: Magma forced up through weak points in the volcano

Effects of the eruption		Response to the eruption	
Primary effects	• About 20 people killed • Two-thirds of homes destroyed and farmland • About 5000 people evacuated	Immediate response	• Evacuation • Emergency help to provide temporary shelters
Secondary effects	• Cost of rebuilding • Tourist industry disrupted	Long-term response	• UK aid to rebuild infrastructure • Volcano Observatory built to monitor volcano • Attempt to reattract tourists

Predicting eruptions

There are ways to monitor changes in volcanoes to help predict an eruption. Two of these are:
- tiltmeters which check for bulges on volcano slopes.
- satellites which monitor for changes in heat activity.

Monitoring is expensive and poorer countries often cannot afford it.

Worked example

Explain why people live near volcanoes despite the dangers of eruptions. **(4 marks)**

Volcanic ash makes soil very fertile and people live near volcanoes because of this. Also, composite volcanoes can often be dormant for centuries, so people do not expect them to erupt.

Now try this

What is the difference between primary effects and secondary effects of a natural hazard? **(2 marks)**

Earthquakes

You need to know about where and why earthquakes happen, their features and how they are measured.

Earthquake facts

Location

Earthquakes happen at all three types of plate margin but are most common on destructive margins.

Features

- **The focus** is the point at which the earthquake happens beneath the ground.
- **The epicentre** is the point on the surface above the focus. This is where the effects of the shock are worst.
- **Shockwaves** radiate out in all directions from the focus.

Measuring earthquakes

There are **two** ways of measuring earthquakes.

- The **Richter** scale is used by seismographs to measure an earthquake's magnitude with values plotted on a logarithmic scale of 1 to 10.
- The **Mercalli scale** indicates the earthquake's impact on people. It isn't based on machine measurements but on human judgement. Total destruction is indicated by the Roman numeral XII (12).

Factors controlling the primary effects of an earthquake

The severity of the primary effects of an earthquake depend on a combination of human and physical factors.

Physical		Physical
• High magnitude on Richter scale • Shallow focus (near the surface) • Sands and clays vibrate more (e.g. Mexico City)	• Low magnitude (below 5) • Focus deep underground • Hard rock surface (e.g. Seattle)	
Great (or total) damage High number of deaths and injuries Mercalli scale VII–XII 7-12	Superficial damage to buildings Few casualties Mercalli scale I–VI 1-6	
• High density of population • Residential area of a city • Self-built housing • Lack of emergency procedures (e.g. Gujarat in India)	• Low density of population • Urban area with open spaces • Earthquake-proof buildings • Regular earthquake drills	
Human		Human

Worked example

tier **H**

Explain why many strong earthquakes occur along destructive plate margins. **(4 marks)**

Tectonic plates that move against each other often get stuck due to friction. In the subduction zones of destructive plate margins, one plate is being forced underneath another, causing energy to build up when they get stuck. So the earthquakes produced when they move again are often very strong.

The global distribution of earthquakes

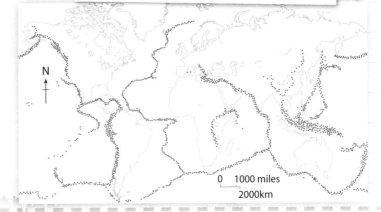

N

0   1000 miles
    2000km

Now try this

tier **H**

 Make two points and develop each one.

Describe the global pattern of earthquakes shown in the map above. **(4 marks)**

Earthquake hazards

You need to know the causes, the effects and the immediate and the long-term responses to an earthquake.

Primary effects

Primary effects are things that happen immediately as a result of an earthquake, for example:

- deaths and injuries
- destruction of buildings or damage to buildings
- destruction or damage to roads, railways, bridges.

How bad primary effects are depends on a mix of physical and human factors, e.g. how strong the earthquake is, whether it happens in a crowded city or a sparsely populated rural area.

Secondary effects

Secondary effects are the after-effects of earthquakes, for example:

- fires caused by fractured gas pipes and broken electricity pylons
- landslides on steep or weak slopes
- spread of disease when sanitation breaks down
- tsunamis, when the earthquake occurs offshore.

Secondary effects may have a bigger impact in poorer countries because they do not have the money to prepare people for earthquakes or protect buildings and infrastructure.

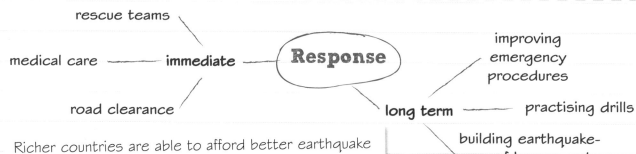

rescue teams

medical care — immediate — **Response**

road clearance

improving emergency procedures

long term — practising drills

building earthquake-proof homes and offices, etc.

Richer countries are able to afford better earthquake preparation and protection than poorer countries.

Worked example

tier **H**

Using an example, outline the impact of a major earthquake on people and property in a poorer part of the world. **(6 marks)**

There was a major earthquake in Haiti in January 2010. Over 316 000 people were killed and many homes were destroyed making over 1 million people homeless. Major buildings were destroyed in the capital Port-au-Prince and roads were badly damaged, so it took over 3 days to reach the smaller towns that were affected. The government found that a further 250 000 homes and 30 000 offices or factories required demolition because they were poorly constructed and unable to withstand the earthquake.

The question asks for the impact on people **and** property, so both must be covered. Make sure you use a named example.

Now try this

tier **H**

Using named examples, compare the responses to an earthquake in a rich and a poor part of the world.

(8 marks)

Tsunamis

Tsunamis are a secondary effect of earthquakes which occur at sea.

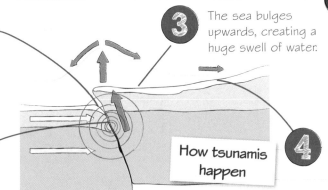

1 Two plates get stuck and build tension.

2 An underwater earthquake happens as the tension breaks, snapping one plate edge upwards.

3 The sea bulges upwards, creating a huge swell of water.

4 The swell reaches the coast and travels very quickly inland.

How tsunamis happen

Case study

You need to know a case study of a tsunami, its causes, effects and responses.

Case study

The Asian tsunami (2004)

The Asian tsunami was caused by an undersea earthquake with an epicentre off the coast of Sumatra, Indonesia. It was one of the most destructive natural disasters ever.

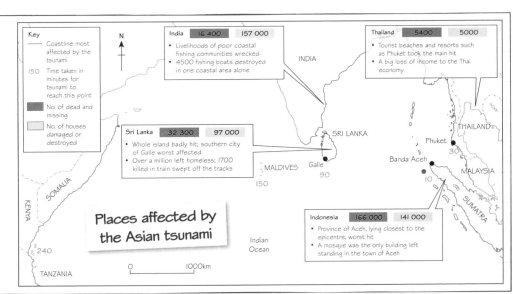

Key
— Coastline most affected by the tsunami
150 Time taken in minutes for tsunami to reach this point
☐ No. of dead and missing
☐ No. of houses damaged or destroyed

India 16 400 157 000
• Livelihoods of poor coastal fishing communities wrecked
• 4500 fishing boats destroyed in one coastal area alone

Thailand 5400 5000
• Tourist beaches and resorts such as Phuket took the main hit
• A big loss of income to the Thai economy

Sri Lanka 32 300 97 000
• Whole island badly hit; southern city of Galle worst affected
• Over a million left homeless; 1700 killed in train swept off the tracks

Indonesia 166 000 141 000
• Province of Aceh, lying closest to the epicentre, worst hit
• A mosque was the only building left standing in the town of Aceh

Places affected by the Asian tsunami

INDIA · SRI LANKA · MALDIVES · Galle · THAILAND · Phuket · Banda Aceh · MALAYSIA · SUMATRA · SOMALIA · KENYA · TANZANIA · Indian Ocean

0 ____ 1000km

The effects of the Asian tsunami

- One of the most destructive natural disasters ever.
- Over 220 000 people killed.
- 2 million people displaced.
- Over 500 000 houses destroyed.
- Infrastructure destroyed or badly damaged.
- Massive economic damage: tourism and fishing disrupted; farmland ruined by salt.

Worked example   tier F&H

Using an example, describe the responses to a tsunami disaster. **(6 marks)**

The Asian tsunami occurred in December 2004. In Banda Aceh, the first area to be hit, hospitals could not cope and bodies were left in corridors. Bodies had to be buried quickly so that disease didn't spread and mass graves were built. International response raised $14 billion, funding the world's biggest ever emergency relief effort. The Disasters Emergency Committee has spent £40 million on rebuilding projects in the area. An Indian Ocean tsunami warning system became operational in 2006.

Now try this

Fill out the case study revision card on page 113 of this book for your tsunami case study.

Rock groups

There are many different types of rock but there are three main rock groups: igneous, sedimentary and metamorphic.

The geological timescale shows when different types of rock were formed

Era Period	Millions of years	Rock type
QUATERNARY	2	Clay
TERTIARY	65	
CRETACEOUS	140	Chalk
JURASSIC	195	Clay
TRIASSIC	230	
PERMIAN	280	
CARBONIFEROUS	345	Carboniferous limestone
DEVONIAN	395	Granite
SILURIAN	445	
ORDOVICIAN	510	
CAMBRIAN	570	
PRE-CAMBRIAN		

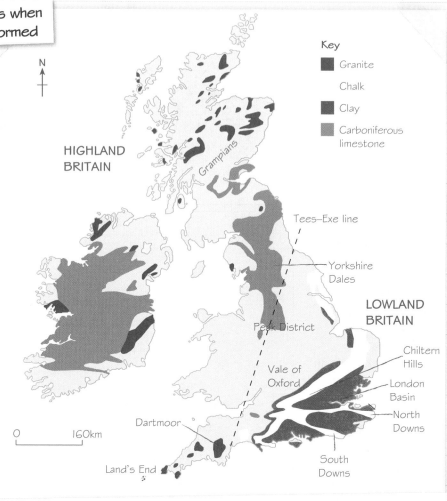

Key
- ■ Granite
- □ Chalk
- ■ Clay
- ▨ Carboniferous limestone

HIGHLAND BRITAIN

Grampians

Tees–Exe line

Yorkshire Dales

LOWLAND BRITAIN

Peak District

Chiltern Hills

London Basin

Vale of Oxford

North Downs

Dartmoor

South Downs

Land's End

0   160km

Some of the rock types of the British Isles
The Tees–Exe line divides the mainly igneous and metamorphic rocks of the north from the mainly sedimentary rocks of the south.

There are **three** groups of rock.

1 **Igneous** – made from magma. For example, basalt and granite.

2 **Sedimentary** – sediment compressed into rock. For example, clay, chalk and limestone.

3 **Metamorphic** – igneous or sedimentary rocks that have been changed by heat and / or pressure. For example, limestone into marble and clay into slate.

Worked example

tier Fn

Choose the correct words to complete the following sentences. **(2 marks)**

**Cretaceous   Jurassic
Carboniferous   Silurian**

The geological era when most limestone was formed is the Carboniferous era.

The Silurian era is an example of a geological era when granite was formed.

Now try this

Describe the process by which sedimentary rocks are formed. **(2 marks)**

tier H

The rock cycle

The rock cycle links how the three rock groups are formed. Weathering is a key stage in the rock cycle. Climate and rock composition influence how much weathering occurs.

The rock cycle

Mountain building (sediments upfolded; pressure from rock crust being destroyed melts the rock into magma which forms volcanoes)

Weathered and eroded materials are deposited on the sea bed, where they are compacted into rock

Rocks are broken down by weathering

Rocks are eroded by water and ice and transported

Weathering and erosion

Weathering is the breakdown of rock at or near the surface. Weathering does not include the transport of the broken-down bits of rock by wind or water – this is called **erosion**.

Mechanical weathering

Freeze–thaw – most common in **cold** climates. When it freezes, water in cracks in the rock expands. Over time the crack widens and pieces of rock fall off. It is most effective when the temperature frequently rises above and falls below 0°C.

Exfoliation – most common in **hot** climates. The surface of rocks heats up, expands, cools down and contracts more than the rest of the rock. Cracks form in and along the top layer and it peels off like the skin of an onion.

Chemical weathering

This happens when the rock's mineral composition is changed.

- Granite contains feldspar which converts to soft clay minerals as a result of a chemical reaction with water.
- Limestone is dissolved by **carbonation**. Carbon dioxide in the atmosphere combines with rainwater to form carbonic acid, which changes calcium carbonate (limestone) into calcium bicarbonate. This is carried away by water in **solution**.

Biological weathering

This is caused by plants and animals and its action speeds up mechanical or chemical weathering. For example, tree roots widen gaps in rocks.

EXAM ALERT!

Some students give a good definition of weathering but forget to say why it is important to the rock cycle.

Students have struggled with exam questions similar to this – **be prepared!**

Worked example

tier F&H

Explain why weathering is an important part of the rock cycle. **(4 marks)**

Weathering breaks down rocks into fragments that can be transported by water and ice. The erosion caused by water and ice is speeded up because they carry the weathered fragments. Chemical erosion also changes rocks that are resistant to erosion into softer materials that are easy to erode.

Now try this

Draw a diagram to show the stages of freeze–thaw weathering. **(4 marks)**

tier H

Granite landscapes

Granite landscapes have distinctive landscape features.

Make your own revision card for this case study based on the example on page 113.

based on the example on page 113.

Case study

You need to know a case study on the ways people make use of a granite landscape.

How granite landscapes are formed	Characteristics	Where it is found in the UK	Features	Land use
Granite forms underground in magma chambers called batholiths. These are exposed over time as rocks above are eroded.	• Hard and resistant to erosion • Impermeable • Irregularly spaced joints • Susceptible to chemical weathering	• North and West of Tees-Exe line • Rare in UK • Areas of high relief	• V – shaped valleys • Tors • Swamps and bogs • Infertile, acidic soils • Moorland	• Grazing for sheep • Reservoirs • Army training • China clay quarrying • Tourism

Case study

Dartmoor – opportunities and limitations

✓ Granite is a good building material so there are many quarries. The impermeable rock and high rainfall make it a good location for reservoirs.

✗ The soil is thin and acidic and not suitable for farming other than sheep grazing.

✗ The weather is often wet and windy.

✓ It became a National Park in 1951. The tors and dramatic scenery are popular with tourists.

Worked example

tier H

Describe **one** theory about how tors are formed. **(4 marks)**

The most widely accepted theory is by Linton (1955). This theory says that when the granite was underground and conditions were warmer and wetter, weathering attacked the granite that had lots of closely spaced vertical cracks more than the granite that had fewer vertical cracks and cracks that were spaced apart more widely. As the granite was exposed on the surface during the Ice Age, the more weathered rock was removed, leaving the less weathered rock as a tor.

Now try this

tier F n

Explain why granite landscapes are not usually very good for agriculture. **(2 marks)**

Carboniferous limestone landscapes

Carboniferous limestone landscapes have many different features.

Make your own revision card for this case study based on the example on page 113.

How Carboniferous limestone landscapes are formed	Characteristics	Where it is found in the UK	Features	Land use
Sedimentary rock formed during Carboniferous period. Landscapes formed when limestone near to surface.	• Hard and strong but with weak joints • Permeable • Many vertical and horizontal joints (bedding planes) • Susceptible to chemical weathering (carbonation)	• Upland areas such as Yorkshire Dales and Peak District • Common in UK	Surface features • Limestone pavements (clints and grykes) • Swallow holes • Dry valleys • Limestone gorges • Resurgence of rivers • Thin soils • Dry Underground features • Caverns • Stalactites and stalagmites • Pillars • Curtains	• Grazing for sheep • Quarrying for building stone and for lime used in chemical processes • Tourism

Tourism in the Yorkshire Dales

The Yorkshire Dales is very popular with tourists and became a National Park in 1954.

✓ Tourism provides jobs for around 13% of working age people in the park.

✓ Tourists spend money when they visit the park: nearly £500 million a year.

✗ Tourism causes problems too, such as erosion of footpaths, congestion on the roads and a high proportion (15%) of houses being bought as holiday homes.

Worked example

tier F&H

Describe the formation of **one** of these underground features of a Carboniferous limestone landscape: **cavern   pillar   curtain**
(2 marks)

Pillar. As water drips from the roof of an underground cave, some of the water evaporates, leaving behind calcium carbonate. Over thousands of years, this builds up into a stalactite at the drip and a stalagmite where it drops. A pillar forms when a stalactite and a stalagmite link up.

Now try this

Explain how dry valleys are formed in Carboniferous limestone landscapes.   **(6 marks)**

Chalk and clay landscapes

Chalk and clay form distinct landscapes.

Make your own revision card for this case study based on the example on page 113.

Case study

You need to know a case study on the way people make use of chalk and clay landscapes.

How chalk landscapes form	Characteristics	Where it is found in the UK	Features	Land use
Chalk is a sedimentary rock formed in the Cretaceous period. Clay is formed by compression of mud deposits. When layers of chalk and clay are tilted up by earth movements to the surface, chalk and clay landscapes form.	Chalk is: • strong with joints and bedding planes • porous and permeable Clay is: • weak • impermeable Springs emerge where permeated water meets the clay layer.	• Chalk is found only in lowland Britain, e.g. North and South Downs • Clay is very common across Britain	Chalk features • Cuesta (chalk escarpments – scarp and dip slope) • Dry valleys and bournes (temporary streams) • High cliffs at the coast which are resistant to erosion • Short, rich turf Clay features • Clay vale – wide, flat plain with lots of drainage • Weak cliffs at the coastline which are vulnerable to erosion • Wet, heavy soils	Chalk • Grazing for sheep • Quarries • Tourism • Water supply (aquifers) Clay • Dairy farming • Brick production

The London Aquifer

London gets its water from an aquifer. Chalk layers store rainwater which cannot escape because of the clay lying below and above it.

Gault clay | Chiltern Hills | Chalk | High Barnet | London clay | Hampstead | River Thames | Sydenham | North Downs | Gault clay | Older rocks

Worked example

Complete the following cross section of a cuesta by labelling the following features:

scarp slope   dip slope   spring line   dry valley
(4 marks)

Spring line | Scarp slope | Dry valley | Dip slope | clay | spring | water table | clay | chalk

Now try this

Explain why settlements are often found along a spring line in chalk and clay landscapes.   (4 marks)

Quarrying

Demand for resources has led to quarrying, but quarrying has advantages and disadvantages.

Advantages

👍 Economic Quarries provide raw materials for buildings, roads and cement manufacture.

👍 Economic Quarries are a major source of employment in areas where often there aren't many other jobs.

👍 Social When quarries close down they are often used for recreation.

Case study

You need to know a case study of the economic, social and environmental advantages and disadvantages of a quarry.

Disadvantages

👎 Environmental Quarries destroy local habitats.

👎 Social / Economic Quarries are dirty and noisy and off-putting to tourists and local residents.

👎 Economic It takes a lot of money to restore the site of a closed quarry.

Case study

Hope Quarry

This quarry is in a Carboniferous limestone landscape that is popular with tourists.

Advantages:
provides 300 jobs in an area where jobs are hard to find.

Disadvantages:
- Unsightly in a tourist area (nearby Castleton is a popular tourist town).
- Noise and air pollution.

Hope Quarry and Cement Works in the Peak District National Park

Management solutions – landscaping:
- 75 000 trees planted to screen site
- rocks placed to hide the entrance
- worked-out pits used as fishing lakes and a golf course.

Quarry controls

Because of the disruption quarries cause, planning permission is only granted for new quarries when they are found to be really necessary. Planners insist that impacts are minimised: the site must be screened; blasting times limited; and the quarry must be reclaimed after the site is worked out.

Be sure to describe the economic benefit of the activity and not just the activity. Also, don't be tempted to 'explain'. The question asks you to 'describe'.

Worked example

tier F&H

Using one of the rock types you have studied, describe the benefits of using this landscape for economic activity. **(4 marks)**

Quarries are often found in Carboniferous limestone areas and they provide jobs for local people in areas where jobs are often scarce. Upland limestone areas like the Peak District are also often very scenic and attract many tourists, who spend money in the local area.

Now try this

Draw a sketch map of your case study quarry. Annotate your map to show the environmental impacts. Use a grid so that you can sketch within it. **(4 marks)**

Quarrying management strategies

The impact of quarrying on the environment can be reduced by careful, sustainable management.

What is sustainable management?

Sustainability is about ensuring the way we use a resource now doesn't hinder the ways future generations can use it. Quarrying is highly unsustainable because quarried material cannot be put back in the quarry for future generations.

It is impossible to make quarrying sustainable but there are ways to reduce its impact, for example, by only allowing blasting at certain times of day and managing the site after quarrying is finished.

Case study

You need to know a case study of a quarry and how it has been managed sustainably to reduce its impact.

Worked example

tier F&H

Some quarries are used as landfill sites once they are worked out. Describe the advantages and disadvantages of this management strategy. **(4 marks)**

One advantage of this strategy is that landfill sites are in short supply and disused quarries are ready-made holes in the ground. One disadvantage is that landfill can sometimes produce toxic run-off which could get into local river systems.

Strategies to reduce the impact of quarries

Old workings can be filled with water to create wetland habitats and areas for recreation.

Trees can be planted to screen off the site for residents and tourists.

Wildlife conservation areas can be established to offset the damage to the local environment when the quarry is worked out.

Worked-out quarries can be reclaimed, filled in and replanted or put to a completely new use, for example as the site for the Eden Project.

Now try this

For your quarry management case study, outline the different strategies that are being used to minimise the environmental impact of the quarry. **(8 marks)**

tier H

The UK climate

The UK has a temperate maritime climate: mild, wet winters and warm, wet summers.
The characteristics of the UK climate can be explained by its position on the globe.

Global influences on UK climate

Distance from the sea: The sea takes longer to heat up than the land so it is cooler in summer. But it takes longer to cool down than the land, so it is warmer in winter. Because the UK is an island, it is very influenced by the sea (maritime) and so does not have any extremes of temperature.

Latitude: The curve of the earth means the equator is hottest and the poles are coolest due to insolation as a result of the angle of the sun's rays. The UK is in the temperate zone of latitude: neither tropically hot nor polar cold.

North Atlantic Drift: This warm current makes the west of the UK warmer and wetter than the east, especially in winter.

Relief: As the air rises over mountains (mostly in the west and middle of the UK) it cools and drops its moisture as rain. This makes the east drier.

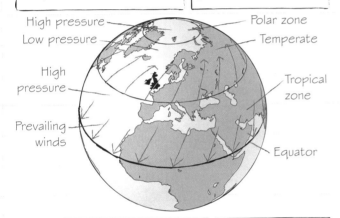

High pressure — Polar zone
Low pressure — Temperate
High pressure —
Prevailing winds —
Tropical zone
Equator

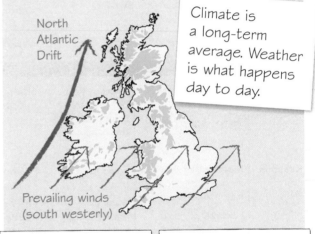

North Atlantic Drift

Prevailing winds (south westerly)

Climate is a long-term average. Weather is what happens day to day.

Winds: Air moves from high pressure zones to low pressure zones. The prevailing winds in the UK are from the south west, over the sea, bringing rain.

Pressure: The UK is in a low pressure zone where air is rising. As it rises it cools and it rains.

Altitude: Temperatures decrease by 1°C for every 100 metres of altitude, so high areas are colder.

Sunshine hours: The far south has the most hours of sunshine a year; the far north has the fewest.

Prevailing winds: These come in from the west and make the west wetter than the east.

Insolation: Summer temperatures are cooler in the north than the south.

Worked example

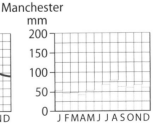

tier **F**n

Study the climate graph for Manchester opposite.

(i) Which month had the highest temperature?

(ii) Which month had the lowest precipitation?

(2 marks)

(i) July.

(ii) April.

Manchester
°C   mm
30 ┤   200
20 ┤   150
10 ┤   100
0   ┤   50
-10 ┤   0
    J FMAMJ JASOND   J FMAMJ JASOND

Remember that the bar graph is the precipitation (usually in mm) and the line graph is the temperature (usually in °C).

Now try this

tier **H**

What is 'insolation' and how does it lead to temperature differences from low to high latitudes? **(2 marks)**

Depressions

Depressions have a big influence on the UK weather. They bring wet and windy weather.

Frontal depressions

A cross section of a frontal depression, moving from west to east, so A is at the start of the depression's passage across the UK. Someone standing at A will experience all the weather features A to E as they pass overhead.

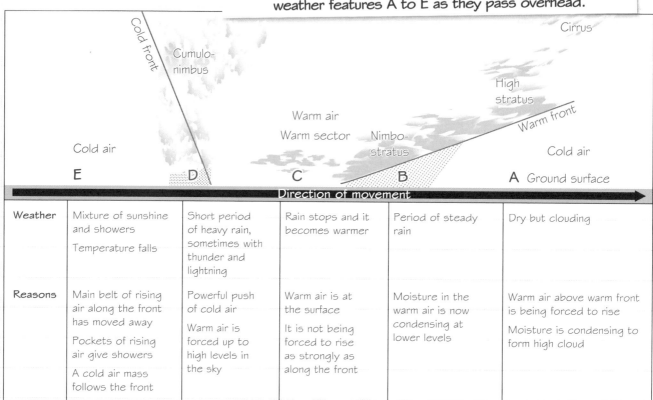

	E	D	C	B	A Ground surface
Weather	Mixture of sunshine and showers Temperature falls	Short period of heavy rain, sometimes with thunder and lightning	Rain stops and it becomes warmer	Period of steady rain	Dry but clouding
Reasons	Main belt of rising air along the front has moved away Pockets of rising air give showers A cold air mass follows the front	Powerful push of cold air Warm air is forced up to high levels in the sky	Warm air is at the surface It is not being forced to rise as strongly as along the front	Moisture in the warm air is now condensing at lower levels	Warm air above warm front is being forced to rise Moisture is condensing to form high cloud

Worked example

tier F n

Study the diagram of weather fronts opposite.
Label the following features on this diagram:
warm front   cold front   occluded front.
(3 marks)

When you are labelling diagrams, make sure you write very clearly and place your labels accurately.

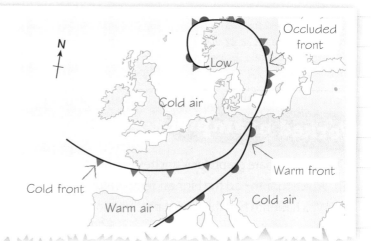

Now try this

tier F n

Complete the gaps in this description of a depression:

Depressions are areas of _____ pressure. In general, depressions bring unsettled weather with wind, cloud and _____. Depressions form where _____ tropical air meets cold polar air.

warm   cold   low   high   fog   rain   isobar   (3 marks)

Anticyclones

Anticyclones also have a big influence on the UK weather. They bring dry weather.

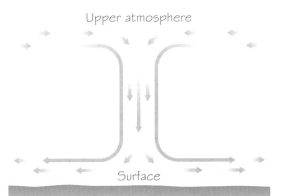

Upper atmosphere

Surface

In an anticyclone, air descends due to **high pressure**. As it falls, it warms up and any moisture it is carrying evaporates.

Features of an anticyclone

- Light winds due to gentle **pressure gradient**.
- Dry air as all the moisture has evaporated.
- Clear skies.
- In summer the sunshine isn't blocked by clouds or cooled by winds so it is hot and dry.
- In winter there is low insolation and heat escapes because there are no clouds, making it cold and frosty.
- 'Anticyclonic gloom' happens in winter: moist air from the sea condenses to form low cloud and fog.
- Sometimes an anticyclone can remain stationary for a while – this is called a blocking anticyclone.

Isobars

- **Isobars** join places with equal atmospheric pressure.
- The numbers show the atmospheric pressure measured in millibars.
- When the highest number is in the middle circle then this shows an anticyclone.
- Isobars that are far apart show calm conditions; the closer together they are, the windier it is.

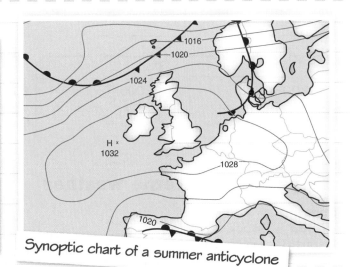

Synoptic chart of a summer anticyclone

This weather map is for 13 September 2012

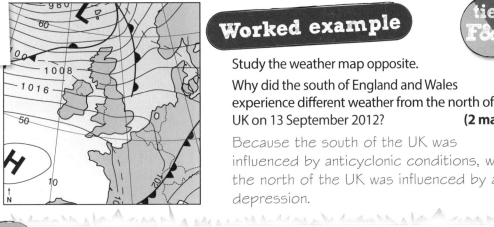

Worked example

tier F&H

Study the weather map opposite.

Why did the south of England and Wales experience different weather from the north of the UK on 13 September 2012? **(2 marks)**

Because the south of the UK was influenced by anticyclonic conditions, while the north of the UK was influenced by a depression.

Now try this

tier H

Explain how a blocking anticyclone can lead to a heatwave in the UK. **(4 marks)**

Extreme weather in the UK

UK weather is becoming more extreme – more severe than what would be expected. This usually has negative impacts on people's lives, but can have positive ones too.

Examples of extreme UK weather

- Big freeze
- Heatwave
- Heavy rain
- Drought
- Gales
- Heavy snowfalls
- Thick fog
- Floods

The UK had a heatwave in 2003

Negative impacts: over 2000 people died in the UK as a result of the heat

Positive impacts: great for UK tourism

Timeline

Evidence of extreme weather events

2004: flash flooding in Boscastle after extremely heavy rain.

2006: dry period (from 2004–2006) for south-east England with below-average rain for 17 out of 21 months.

2008: record low temperature for October.

2011: exceptionally warm and dry spring.

2003: heatwave. A new high temperature record of 38.5°C was set in August in Kent.

2005: flooding in Carlisle and North Yorkshire.

2007: flooding in many UK areas, especially Gloucestershire following record rain levels for July.

2010: Big Freeze in January.

2012: wettest year in England since records began.

Impacts of extreme weather

crops destroyed on farms

loss of stock, e.g. in shops

loss of possessions

value of homes drop

businesses flooded

homes flooded

unable to get insurance

Effects of flooding on people's lives

cost of repairs

health

transport routes disrupted

damp conditions – unhealthy

flooded drains a health hazard

people can't get to work

children can't get to school

Worked example

tier H

Describe ways in which the impacts of extreme weather events can be reduced. **(4 marks)**

The key to tackling extreme weather events is improving prediction. People can be warned in time for them to take action, be prepared, and be ready to deal with an extreme event (for example, having sandbags ready to use against flooding). They can also develop plans for dealing with the impact of an extreme weather event (for example, clean water supplies in case of flooding).

Now try this

Explain why some people think the increased frequency of extreme weather events in the UK could be linked to climate change. **(4 marks)**

tier H

Global warming

The majority of scientists agree that the Earth's climate is warming and that the warming is caused by human activity. There are also natural causes of temperature change.

Natural causes of global temperature changes

- Orbit wobbles – Earth's orbit changes slightly once per 100 000 years (called Milankovitch cycles).
- Sun output – the Sun gives out more or less radiated heat energy over an 11-year cycle.
- Ocean currents – for example, in the UK warm Atlantic currents keep our climate fairly warm. If these currents shifted further away, this would affect our climate.

Human causes of global warming

Global warming is closely associated with rising atmospheric CO_2. It is a powerful greenhouse gas released by many human activities, including:

- energy production
- transport
- industry
- deforestation.

None of these are able to account for the degree of warming that is occurring.

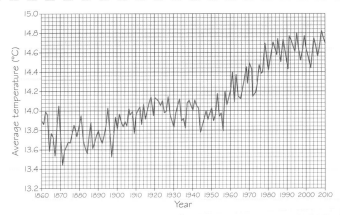

Average global temperature increase, 1860–2010

- On average, global temperatures are 1°C higher now than at the end of the 19th century.
- The 1990s saw some of the highest temperatures ever recorded.
- Some experts predict a rise of 4.5°C by the end of the 21st century, which would have catastrophic consequences.
- Climate change sceptics may agree with the existing data but disagree about the causes and about the forecasts.

How does the greenhouse effect work?

The greenhouse effect is a natural process. We need greenhouses gases – just not too much!

 Sun shines onto the Earth and warms the surface.

 Warmth radiates up from the surface and into the atmosphere.

3 Greenhouse gases prevent some of the warmth escaping back into space and warmth is trapped (like the glass in a greenhouse).

Worked example   tier H

Explain the difference between climate change and global warming. **(4 marks)**

Climate change has happened many times over the course of the Earth's history, driven by natural factors such as changes in the Earth's orbit of the Sun. While these changes have involved warming and cooling of average global temperatures, global warming is used to describe the current phase of rapid increase in average global temperatures, which many scientists believe is due to human activity.

Now try this

Which of the following types of human activity is **not** a major source of greenhouse gases? **(1 mark)**

Cement making   Steel production   Solar power generation   Car manufacturing

 tier Fn

Consequences of global climate change

Future climate change could have major environmental, economic, social and political consequences around the world – including the UK. This could change the way we live.

Worldwide effects of global warming

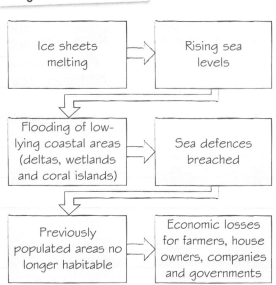

Ice sheets melting → Rising sea levels

Flooding of low-lying coastal areas (deltas, wetlands and coral islands) → Sea defences breached

Previously populated areas no longer habitable → Economic losses for farmers, house owners, companies and governments

Worked example

tier H

Explain some of the possible **economic impacts** of climate change on the UK. **(4 marks)**

Climate change in the UK could cause flooding in London. This would be very serious as many of the UK's most important businesses are in London. Another problem is probably less reliable rainfall and therefore drought. It would cost money to move water around different parts of the country.

If a question uses the plural, e.g. 'impacts' rather than 'impact' as in this question, make sure you mention more than one.

Environmental Sea level rise would mean loss of coastal land and more erosion with a risk of flooding of low-lying cities, e.g. London.

There may be more severe storms and longer summer droughts.

Ecosystem change could mean some plant and animal species move into new areas, new (invasive) species appear.

Political Refugees from countries hit hard by climate change may come to the UK.

The cost of protecting coasts and cities from flooding will be unaffordable in some cases.

Possible consequences of future climate change in the UK

Economic Damage to cities like London from flooding would be extremely expensive and very disruptive. Insurance premiums for people in flood risk areas would increase.

Warmer weather might mean farmers can grow different crops and enjoy longer growing seasons.

Social Hotter summers could mean more people have holidays in the UK.

Warmer temperatures could see diseases like malaria come to the UK.

Now try this

tier H

Explain how climate change could lead to an increase in conflicts around the world. **(6 marks)**

Responses to the climate change threat

The threat of global climate change can only be combated by a united international response, with action at international, national and local levels. How is this progressing?

Ways to reduce greenhouse gas emissions:
- Reduce the use of fossil fuels – for example, by fining industries for high emission levels.
- Find alternative energy sources.
- Reduce deforestation and increase afforestation.
- Develop carbon capture technologies.

International cooperation

- Kyoto Protocol (1997): global agreement to cut emissions of CO_2 by 2012. 37 industrialised countries signed the protocol but few met their targets.
- Copenhagen Accord (2009): a new agreement that softened terms of Kyoto. Countries were asked to say what cuts they could make to CO_2 emissions by 2020.

International
- Kyoto Protocol
- Copenhagen Accord
- Carbon credits

National
- CO_2 emission reduction targets
- Improve public transport
- Schemes to reduce energy consumption, e.g. high road taxes for 'gas guzzlers'

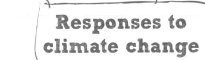

Responses to climate change

Local
- Congestion charging
- Recycling services
- Energy efficiency schemes for homes and businesses

Individual
- Reduce consumption and energy use
- Reuse more products
- Recycle

Kyoto also set up a Clean Development Mechanism (CDM):
- Countries that beat their emissions targets get **carbon credits** which they can sell to other countries.
- Countries that help poor countries beat emissions targets also get credits.

Worked example

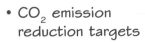

Describe how congestion charge schemes aim to reduce carbon emissions. **(2 marks)**

By discouraging people from bringing their cars into the congestion charge zone. By encouraging people to use public transport instead, which means lower carbon emissions.

Now try this

'Outline' means give the most important points.

Outline the terms of the Kyoto Protocol for industrialised countries. **(2 marks)**

Causes of tropical revolving storms

Tropical revolving storms are a major climate hazard which affect areas such as the Gulf of Mexico, the Caribbean, India and China.

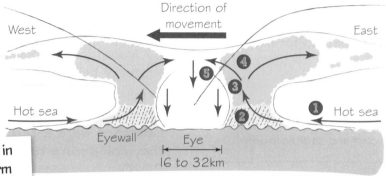

The eyewall surrounds the eye. Here the air is rapidly spiralling upwards and there are high winds and torrential rain.

Direction of movement

West

East

The eye is the centre of the storm. Here there is falling air, light winds and no rain.

Hot sea

Hot sea

Eyewall

Eye
16 to 32km

What happens in a tropical storm

Tropical storms need certain **conditions** to develop.

- They need warm water over 26.5°C. This is why they occur in late summer or autumn when the sea has had time to warm up.
- Latitude needs to be greater than 5° north or south. Closer to the equator there isn't enough spin from the Earth's rotation (**Coriolis effect**).
- Winds need to be the same strength from ground level to 12 km high. Different wind speeds pull the storm apart.

Tropical storms **develop** in **five** main stages.

1. Warm, moist air rises and condenses, releasing huge amounts of energy, powering the storm.
2. As the air rises up, it sucks in more warm, moist air behind it.
3. The air spirals upwards rapidly which causes high winds.
4. The air condenses as it rises and cools, forming huge clouds and heavy rain.
5. In the centre of the storm, air falls, forming the eye.

Tropical storms:

- move westwards because winds blow from the east near the equator.
- lose strength on land because they no longer have warm seawater to power them.
- occur in the tropics.

Worked example

What are the names of tropical revolving storms in different parts of the world?   **(2 marks)**

Hurricanes in North America, typhoons in East Asia and Japan, cyclones in the Indian Ocean, Willy Willies in Australia.

tier **F&H**

Now try this

tier **F n**

Describe **two** hazards that can occur when tropical revolving storms reach inhabited coastlines?   **(4 marks)**

Comparing tropical revolving storms

Case study

You need a to know a case study of a tropical storm in a rich country and a poor country

The effects of and responses to tropical revolving storms differ between richer and poorer areas of the world because rich countries can afford better prediction, preparation and protection.

Case study

Poor country

Date: 2 May 2008

Name and location: Cyclone Nargis, Myanmar

Storm details: Winds up to 215 km/h, high winds, heavy rain and storm surges. Irrawaddy Delta region worst affected.

Effects

Social at least 140000 dead, up to 3 million homeless, 95% of homes destroyed.

Economic $10 billion needed to rebuild homes and clear transport routes.

Environmental drinking water polluted, diseases spread, farmland damaged by salt.

Responses

- Short-term: UN launched a huge appeal but the government was reluctant to let foreign organisations help.
- Long-term: relying on foreign aid to rebuild; slow recovery.

Case study

Rich country

Date: 29 August 2005

Name and location: Hurricane Katrina, USA

Storm details: Category 3 hurricane, winds up to 280 km/h, 8.5 m high storm surge punched through levee defences of New Orleans causing widespread flooding.

Effects

Social 1836 dead, 1 million homeless, lack of clean water and sanitation, problems with looting and disorder.

Economic $98 billion to rebuild housing and to care for homeless and jobless.

Environmental coastal habitats damaged, floodwaters carried pollution.

Responses

- Short-term: massive rescue effort and rush to provide shelter and support.
- Long-term: major rebuilding and improvement of levées and improvements to monitoring and forecasting.

Worked example

tier **H**

Compare how prediction, protection and preparation differed in the responses to a **named** tropical revolving storm in a rich and a poor part of the world.

(8 marks)

Prediction: Katrina was tracked and people given plenty of warning. Cyclone Nargis was tracked by Bangladesh, but the Myanmar government did not act on the information.

Protection: Katrina's storm surges were too powerful for New Orleans' levee system but this has since been upgraded. Myanmar cannot afford the huge investment required for protection. Both countries had reduced natural protection by clearing coastal marshes.

Preparation: 80% of people were evacuated from New Orleans; most people in Myanmar had no idea Nargis was coming.

This answer needs more relevant detail, e.g. the different ways data was obtained on how Katrina was developing or the cost of the upgrading of the levée system.

Now try this

Use the case study revision card on page 113 of this book to create revision cards for your two tropical storm case studies.

What is an ecosystem?

An ecosystem is made up of plants and animals and the physical factors that affect them (climate and soil). These different parts are interrelated and depend on each other. A change in one part of the ecosystem impacts on the other components.

A hedgerow: a small scale ecosystem

The wet climate and rich soils of the UK mean a wide variety of plants grow in a hedge supporting many different organisms.

A hedgerow

home to animals, birds and insects

climate is warm in summer, wet and cool in winter

hedgerow plants

fertile soils enriched by rotting vegetation

Food chains and food webs

A food chain

Producers use energy from the Sun to make food

Consumers get their energy from eating producers or ...

Some consumers eat other consumers to get their energy

In an ecosystem the flow of energy and the cycling of nutrients need to be constant to maintain the balance. Nutrients are recycled with the help of **decomposers** which break down the dead remains of plants and animals and release chemicals for plants to use again.

A **food chain** describes the flow of energy from a producer to a herbivore (plant consumer) and on to a carnivore (herbivore consumer). At each stage, less energy is passed on.

A **food web** illustrates all the different food chains in an ecosystem and how they connect together. In a small-scale ecosystem such as an oak tree, aphids and caterpillars feed on the leaves, spiders eat aphids, blue tits eat spiders and caterpillars, and sparrowhawks eat blue tits.

Questions like this are looking for a definition only.

Worked example

tier F&H

A simple food chain on this page shows that rabbits eat plants and foxes eat rabbits. Suggest what might happen if there was a sudden change to the number of rabbits. **(4 marks)**

If the number of rabbits dropped suddenly, plants such as grasses would grow more. But the number of foxes might drop if rabbits were what they mostly ate. However, if the number of rabbits increased rapidly, there might be a growth in the fox population for a while. But, then the rabbits might eat all the grass, rabbit numbers would crash and then fox numbers would too.

Now try this

What is meant by the term **nutrient cycling**?　　　　(1 mark)

tier F&H

Ecosystems

Different ecosystems are found in different parts of the world because of the influences of different climates and different soils.

Tropical rainforest
- Hot all year
  Average temperature 27–30°C
- Wet all year
  Annual precipitation 2000–3000 mm

Temperate deciduous forest
- Warm summers
  Average temperature 16–20°C
- Mild/cool winters
  Average temperature 3–8°C
- Precipitation all year
- Annual rainfall 550–1500 mm

Hot desert
- Very hot in summer
  Average temperature 35–45°C
- Hot in winter (20–30°C)
- Dry all year
- Annual rainfall under 250 mm.

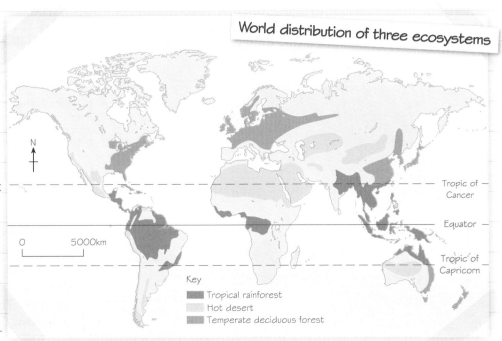

World distribution of three ecosystems

Tropic of Cancer

Equator

Tropic of Capricorn

0   5000km

Key
- Tropical rainforest
- Hot desert
- Temperate deciduous forest

Tropical rainforest ecosystems are very rich in plant and animal life. Soils are thin as nutrients are cycled quickly before the heavy rain washes them away. Plants are adapted to heavy rainfall (waxy leaves with drip-tips) and trees grow tall in competition to reach the sunlight (buttress roots for support). There are three distinct layers to the rainforest structure.

Temperate deciduous forest trees drop their leaves as an adaption to winter conditions (colder, shorter days). The leaves build up into deep, brown earth soils. In spring, ground-level flowers such as bluebells come out before the tree leaves block too much of the sunlight.

Worked example

tier F n

The photo opposite shows a hot desert ecosystem. Label the photo to briefly describe the soil and vegetation cover.　**(4 marks)**

vegetation: plants are scattered to reduce competition for water and have deep roots

soil: very poor, almost no organic content

Now try this

Describe the global distribution of tropical rainforest ecosystems.　**(2 marks)**

tier F&H

Temperate deciduous woodlands

Woodland in the UK is largely made up of deciduous trees. These woodlands need to be used and managed sustainably.

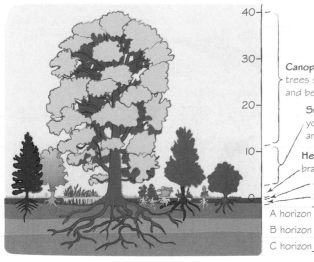

40 —
30 —
20 —
10 —
0

Canopy layer: trees such as ash, oak and beech

Sub-canopy layer: younger trees and tall shrubs

Herb layer: bramble, bracken, bluebells

Ground layer
Litter

A horizon
B horizon
C horizon
Soil: typically brown earth

Case study

You need to know a case study of a temperate deciduous woodland.

- Deciduous woodland is used for timber, recreation and conservation.
- Soils are rich and trees can be felled and new ones planted without soil nutrients being washed out.
- The canopy blocks some light, but not so much that the forest is dark. The forests are very pleasant places for recreation.

controlled felling when cover too dense and for commercial purposes

replanting for commercial use and by sponsorship — trees

Ways to manage woodland sustainably

cycle tracks, footpaths etc.

tourism and recreation

visitor centres

education programmes

conservation of woodland

policing, monitoring and fines

Case study

The National Forest

The National Forest covers 520 km² in the English Midlands and was created in 1990 to help restore woodland to this area. Wooded cover has been increased from 6% to nearly 20% through the planting of over 7 million trees. The overall aim is to increase the woodland cover to a third of the total area, with 30 million trees. This will benefit the local economy, provide recreation facilities and maintain natural habitats.

Worked example

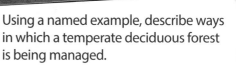

Using a named example, describe ways in which a temperate deciduous forest is being managed.   **(4 marks)**

The National Forest is being managed in different ways to benefit different groups. Commercial forestry thins out overcrowded planting and removes old trees in a way that makes money without damaging the ecosystem. Recreational management sets out trails and facilities so people can enjoy the forest environment in a controlled way.

Now try this

What is meant by 'sustainable management' of woodland ecosystems?   **(2 marks)**

Deforestation

Deforestation affects many tropical rainforests. This has serious social, political and environmental impacts.

The structure of a tropical rainforest

The trees compete for light and are adapted to conditions.

A   Tallest trees (called emergents)

B   Continuous layer of the main canopy

C   Discontinuous under-canopy of trees between 10m and 20m high

D   Layer of shrubs and young trees

E   Herb layer with ferns 6m or more high

Metres
40 —
30 —
20 —
10 —
0 —

The biodiversity of the rainforest makes it a unique ecosystem supporting 50% of the world's 10 million species.

Traditional farming (slash and burn)

Commercial farming and ranching

Logging

Reasons for deforestation

Road building

Population pressure

Mineral extraction

Worked example

tier **Fn**

The diagram opposite identifies different causes of rainforest deforestation. Choose **one** and outline how it leads to deforestation. **(2 marks)**

Road building increases deforestation because not only are trees removed to make the road but roads then open up forest areas to commercial activities such as farming, settling, mining and logging.

Impacts

Environmental
- Soil erosion because the soil is no longer protected by roots and vegetation
- Increased CO_2 released when trees burned.
- Reduces biodiversity.
- Fewer trees to store CO_2 and produce O_2.

Social
- Conflict over use of forest by different interests.
- Development may mean more social facilities for locals.

Economic
- Jobs for local people.
- Big profits for big companies.
- Provides places for migrants to set up homes and farms.

Political
- International pressure to conserve rainforest.

Now try this

This question is looking for a definition **and** a developed point.

What is meant by biodiversity and how does deforestation reduce it?   **(3 marks)**

tier **H**

Sustainable management of tropical rainforest

Sustainable management means using resources now in such a way that future generations will still be able to use the same resources to meet their needs.

Strategies

- Set up national parks and reserves to control development within the forest
- Control logging by restricting logging licences and protecting valuable trees
- Replant felled areas
- Practise agroforestry – combining crops and trees together on the same land (either growing crops amongst the forest or growing trees on cropland)
- Find alternatives for the resources being taken out of the rainforest and develop alternative energy sources to reduce reliance on wood for fuel
- Encourage sustainable eco-tourism that benefits local people without damaging the rainforest
- Educate local people about the benefits of protecting the forest
- Introduce debt schemes where debt to richer countries can be swapped for rainforest conservation

Eco-tourism

A rainforest

International cooperation

- The International Tropical Timber Agreement (ITTA) (2006) aims to promote sustainable forest management.
- The Forest Stewardship Council encourages customers to buy wood from sustainable sources.
- International pressure groups such as Greenpeace monitor deforestation and help track down illegal loggers.

Worked example

tier F&H

Explain how ecotourism provides a sustainable way of making use of tropical rainforests. **(4 marks)**

Ecotourism aims to benefit local people without damaging the forest ecosystem. Tourist numbers are kept low, facilities reuse and recycle as much as possible and local people are employed as staff, guides and educators in a way that values traditional knowledge and skills.

EXAM ALERT!

Some students fall into the trap of 'describing' rather than 'explaining'. You must show **how** eco-tourism can be a sustainable strategy for use in tropical rainforest management.

Students have struggled with exam questions similar to this – **be prepared!**

Now try this

tier H

Suggest ways in which reducing international debt could help protect rainforest resources. **(4 marks)**

Rainforest management

The Amazon rainforest is the largest rainforest in the world. Since 1970, 230000 square miles have been cleared, mostly through **deforestation**.

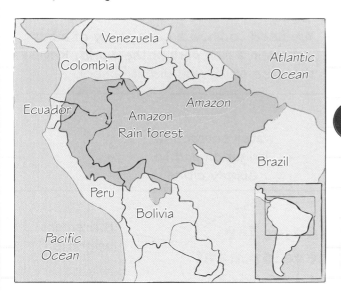

Venezuela
Colombia
Atlantic Ocean
Ecuador
Amazon
Amazon Rain forest
Peru
Bolivia
Brazil
Pacific Ocean

Case study

You need to know a case study of a tropical rainforest, causes and impacts of deforestation and methods of sustainable management. Make sure you revise the case study you did at school.

Worked example

tier **Fn**

Give **two** reasons why tropical rainforest deforestation is a global problem. **(2 marks)**

Burning trees to clear land means a lot of carbon is released, increasing global warming. The rainforest is a carbon sink that the whole world benefits from.

Case study

Amazon rainforest

Causes of deforestation

- Mining
- Road building
- Logging for timber exports
- Huge cattle ranches and other large-scale agricultural uses
- Reservoirs and dams for HEP schemes

Reducing deforestation

satellite surveillance to monitor any deforestation

committed to reducing deforestation

Brazilian government policies

crack down on illegal logging

promotion of sustainable rainforest management

Impacts of deforestation

There are many impacts; here are four.

Economic 70% of deforested land in Brazil is used for cattle ranching. Beef exports were worth $5.37 billion in 2011.

Economic Poor farmers can make money by selling trees on their land to illegal loggers.

Social Native tribes, such as the Guarani people, forced to move to make way for farming.

Environmental Three-quarters of Brazil's total carbon emissions are due to deforestation.

Use the case study revision card on page 114 of this book to make sure you have all the detail you need.

Now try this

Your answer should refer to the environmental impact of deforestation on tropical rainforest soils.

tier **F&H**

Using your knowledge of the impacts of rainforest deforestation, suggest a reason why Brazilian cattle ranchers often have to abandon pasture land every few years and clear new land from the forest. **(2 marks)**

Economic opportunities in hot deserts 1

There are economic opportunities in deserts but even in rich countries water is a major issue.

Cotton plants need irrigation to grow in the Arizona desert

water sources for irrigation and to supply cities

commercial farming

Opportunities for economic development of a desert in a rich country

retirement migration

mineral extraction

tourism

This copper mine in Utah is one of the largest in the world

The Grand Canyon in Arizona attracts around 4.5 million visitors each year

Desert challenges

- **Water:** for example, the Colorado River runs through the desert in western USA and is the main water source for cities in the area which are expanding rapidly, e.g. Las Vegas. The river is struggling to meet demand.
- **Tourism:** the fragile desert environment needs to be protected against damage from rising tourist numbers.

Worked example

tier F n

Give **two** examples of how sustainable management of a hot desert area is helping to reduce challenges faced.   **(2 marks)**

National Parks protect the desert environment in the Mojave Desert. Las Vegas council encourages families to replace water-hungry lawns with dry gardens that use much less water.

You could also **list** two examples, as the command word is 'give'.

Now try this

Desert locations in rich countries are often popular with older people looking for somewhere to spend their retirement. Suggest **one** reason to explain this.   **(1 mark)**

tier F n

Economic opportunities in hot deserts 2

There are economic opportunities in hot desert areas in poorer countries but there are also challenges.

Traditional farming in the Sahara was based on nomadic livestock herding. Now population pressure often means that livestock overgraze the vegetation, leading to soil erosion.

farming

irrigation

Irrigation, like in these fields in Egypt, means crops can be grown in hot desert regions. But overuse of irrigation can cause salinisation, which is the build-up of salt in the soil.

energy

Opportunities for economic development of desert in a poor country

This oilfield is in Algeria; Algeria is also developing solar power stations in the desert.

hunting and gathering

Desert challenges

Population growth

Increase in sheep/goats → overgrazing → soil erosion

Increase in crop cultivation → over watering → salinisation

tier H

Worked example

Give **two** examples of how sustainable management of a hot desert area is helping to reduce challenges faced. **(2 marks)**

Land can be protected from soil erosion by building small stone or earth walls across fields to reduce surface run-off. Growing crops as well as raising animals means fewer animals need to be kept and crops help stabilise the soil.

Now try this   **tier F&H**

Draw a diagram to explain the salinisation process.   **(4 marks)**

Diagrams need to be clear and well-labelled but don't spend lots of time making them look good.

Changes in the river valley

Rivers, and the valleys they flow in, change in different ways between their source (where they start) and their mouth (where they join the sea).

Cross profile and long profile of a river

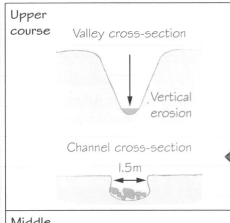

Upper course
Valley cross-section
Vertical erosion
Channel cross-section
1.5m

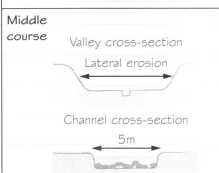

Middle course
Valley cross-section
Lateral erosion
Channel cross-section
5m

Lower course
Valley cross-section
Flood plain
Channel cross-section
20m

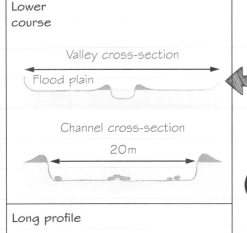

Long profile

The shape of a river valley changes as the river flows downstream. This depends on whether **erosion** is dominant or deposition is dominant. This usually depends on the river's energy.

In the upper course, the river has lots of energy. It is trying to reach its base level, far below. It mostly erodes downwards. This is called vertical erosion. Vertical erosion in the upper course produces a V-shaped valley.

In the middle course, the river is closer to its base level. Vertical erosion is less. The river uses a lot of energy to **transport** its load of eroded material.

In the middle course, the river uses surplus energy to erode sideways. This is called lateral erosion. The valley is wider and flatter and the slopes are more gentle.

In the lower course, the river is close to its base and carries a heavy load of eroded material. If the river slows down, it deposits material.

Worked example

tier Fn

Name three different ways a river and its valley change from source to mouth. **(3 marks)**

Upper course: steep-sided, V-shaped valleys. Middle course: wider channel and gentler valley slopes. Lower course: floodplain, very low gradient as the river gets closer to the sea.

Now try this

tier F&H

Explain the difference between a cross profile of a river and a long profile of a river. **(2 marks)**

Erosion and transportation

River valley shape is determined by the amount of erosion and deposition occurring.

The four main processes of **erosion**

Hydraulic action
The force of the water on the bed and banks of the river removes material

Attrition
The load that is carried by the river bumps together and wears down into smaller, smoother pieces

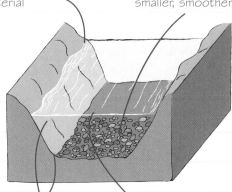

Abrasion
Material carried by the river rubs against the bed and banks and wears them away

Solution
Some rock minerals dissolve in river water (e.g. calcium carbonate in limestone)

Worked example
tier **H**

Suggest why hydraulic action is especially powerful when a river is in flood. **(2 marks)**

Because the river is flowing very fast with a great deal of energy, the force of the water that powers hydraulic erosion will be very strong.

EXAM ALERT!

Some students just describe what hydraulic action is – make sure you answer the question asked.

> Students have struggled with exam questions similar to this – **be prepared!**

The four main types of **transportation**. Transportation is the way in which the river carries eroded material.

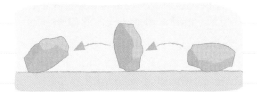

Traction: large boulders roll along the river bed

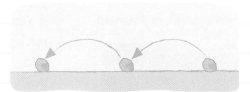

Saltation: smaller pebbles are bounced along the river bed, picked up and then dropped as the flow of the river changes

Suspension: finer sand and silt particles are carried along in the flow, giving the river a brown appearance

Solution: minerals, such as limestone and chalk, are dissolved in the water and carried along in the flow, although they cannot be seen

When the river loses energy (slows down) it may drop some of its load. This is called **deposition**.

See p. 37

For more on depositition.

Now try this
tier **F**

Suggest **three** reasons why a river might slow down enough to deposit material. **(3 marks)**

Waterfalls and gorges

Waterfalls and gorges are formed by erosion in the upper course of the river.

Waterfalls

A waterfall is a steep drop in a river's course.
The diagram explains how they are formed.

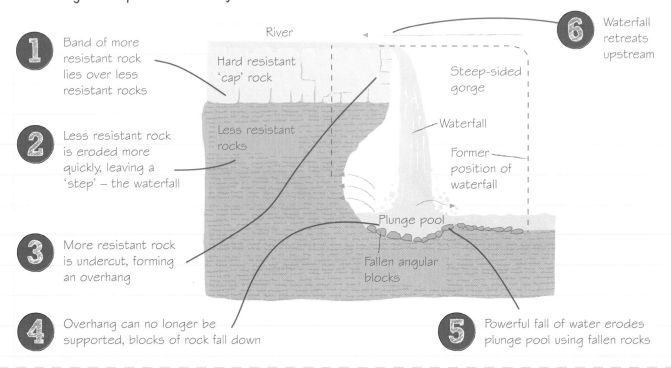

1 Band of more resistant rock lies over less resistant rocks

2 Less resistant rock is eroded more quickly, leaving a 'step' – the waterfall

3 More resistant rock is undercut, forming an overhang

4 Overhang can no longer be supported, blocks of rock fall down

5 Powerful fall of water erodes plunge pool using fallen rocks

6 Waterfall retreats upstream

River

Hard resistant 'cap' rock

Less resistant rocks

Steep-sided gorge

Waterfall

Former position of waterfall

Plunge pool

Fallen angular blocks

Gorges

Over a very long time, the process of undercutting and collapse is repeated and repeated, and the waterfall retreats upstream. A steep-sided gorge is formed.

High Force waterfall on the River Tees

Worked example

tier **H**

What **two** types of erosion are usually dominant in the formation of a waterfall plunge pool? **(2 marks)**

Hydraulic action from the force of the waterfall hitting the bed of the plunge pool, and abrasion of the bed and sides of the pool caused by the rocks that have fallen down from the eroded hard cap of the waterfall.

Now try this

A river's course means upper course, middle course or lower course.

tier **F n**

1 Where in a river's course are waterfalls usually found? **(1 mark)**

2 Why are waterfalls often associated with steep-sided river gorges? **(2 marks)**

tier **F&H**

Erosion and deposition

Meanders and ox-bow lakes and floodplains are landforms created by lateral erosion and deposition form in the middle and lower courses of a river. Levées are landforms created by deposition.

Meanders

On the inside of the bend the current is slower = deposition

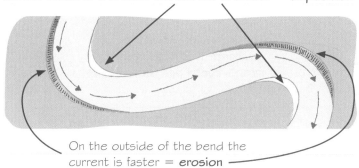

On the outside of the bend the current is faster = **erosion**

> **Meanders** are bends in the river's course. In the lower course, the river uses up surplus energy by swinging one way and the other, causing lateral erosion on the outside of bends and deposition on the inside.

Floodplains are flat areas of land along a river. Lateral erosion makes the valley wider. The river deposits silt on the floodplain each time it floods.

Ox-bow lakes

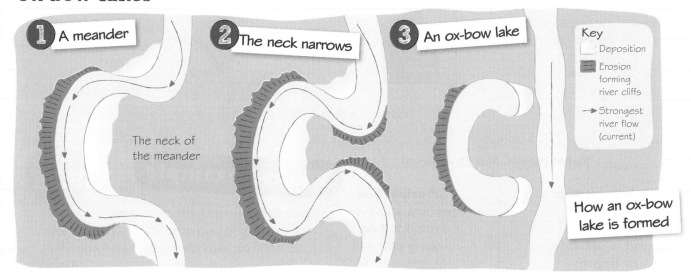

① A meander

The neck of the meander

② The neck narrows

③ An ox-bow lake

Key
- ☐ Deposition
- ▨ Erosion forming river cliffs
- → Strongest river flow (current)

How an ox-bow lake is formed

Levées

As the river floods over its bank, the water slows down. The water can't carry the biggest and heaviest silt particles and they are dropped straight away on the bank forming floodplains.

Increased deposition on the river bed when the river is low gradually raises the river bed upwards

After many floods, the deposits on the bank build up, forming levées.

Worked example　tier H

What is a levée?　**(2 marks)**

A natural embankment along the banks of a river.

Now try this　tier Fn

Explain how both erosion and deposition are involved in the **formation** of a floodplain.

(6 marks)

In any question asking for an explanation of the formation of a landform, make sure the sequence is **complete** and in the right order. Also show that you understand the erosional and depositional processes involved.

Flooding 1

River flooding occurs when the volume of water discharge is so great that all the water cannot be contained in the river channel.

A flood hydrograph (or storm hydrograph) shows how a river responds to a rainstorm.

The lag time is the difference in time between the peak of the rainstorm and the peak of the river discharge.

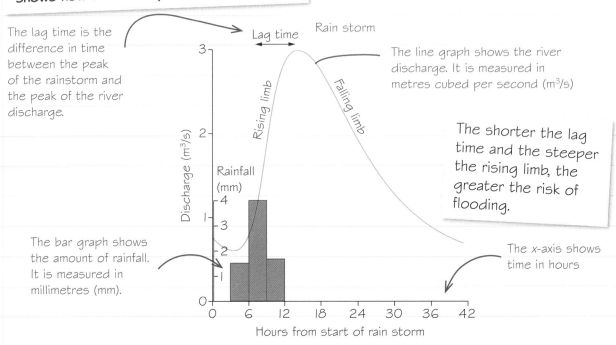

The line graph shows the river discharge. It is measured in metres cubed per second (m³/s)

The shorter the lag time and the steeper the rising limb, the greater the risk of flooding.

The bar graph shows the amount of rainfall. It is measured in millimetres (mm).

The x-axis shows time in hours

Relief (height and steepness)

Previous weather conditions (whether the ground is saturated or frozen)

Precipitation (amount and type: prolonged rain, heavy rain and snowmelt)

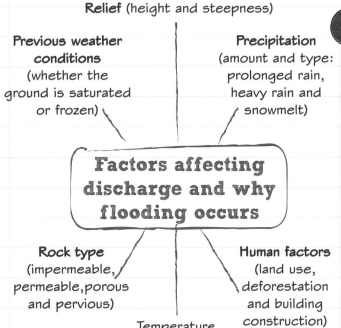

Factors affecting discharge and why flooding occurs

Rock type (impermeable, permeable, porous and pervious)

Human factors (land use, deforestation and building construction)

Temperature (low temperature means less evaporation)

Worked example

tier F&H

Study the diagram opposite showing factors affecting discharge. Choose **two** of these factors and explain how each can make flooding more likely. **(4 marks)**

Flooding occurs when discharge becomes greater than the river channel can contain.

Steep slopes (relief) cause fast surface run-off after heavy rain, which means more water reaches the river more quickly than would happen with gentle slopes.

Land cleared of trees and other vegetation (e.g. for farming) reduces interception, so more rainwater drains into the river than would be the case if the slopes were heavily vegetated.

Now try this

You need to know about the frequency and location of flood events in the UK in the last 20 years so you can answer questions about whether flooding is happening more often. List the main flood events of the last 20 years and say where they occurred.

tier F&H

Flooding 2

You need to know causes, effects and responses to flooding in different parts of the world.

Case study

You need to know a case study of flooding in a rich area and a poorer area of the world.

Case study

Flooding in a rich area

Cumbria, November 2009

Causes
- 31.4 cm of rain in 24 hrs
- ground already saturated
- Derwent and Cocker rivers already swollen

Effect
- 1 person killed
- many people evacuated; 1300 homes were flooded
- 4 bridges collapsed and others were closed due to flood damage
- cost of repair: over £100m

Response
- 200 people rescued from their homes by emergency services
- the army built a temporary footbridge in Workington
- Cumbria Flood Recovery Fund raised £1m to help victims
- the UK government gave £1m for repairs and clean-up

Case study

Flooding in a poorer area

Pakistan, July 2010

Causes
- unusually heavy monsoon rains
- rivers burst their banks all through Indus river basin – 69000 km² affected

Effect
- at least 1600 people killed
- 20 million people affected, 700000 homes flooded
- 1000 bridges washed away, 6.5 million acres of crops washed away
- cost of repair: $4 billion

Response
- Pakistan government criticised for slow response
- Pakistan military involved in rescue effort, rescuing 350000 people
- international organisations raised $687m
- 3 months later, 7 million people were still without adequate shelter

Worked example

Use one effect from the diagram opposite to compare the impact of named flood events in a poorer and a richer area of the world. **(4 marks)**

Disease and illness. In the 2010 floods in Pakistan, millions of people were exposed to gastroenteritis and diarrhoea as a result of contaminated water supplies. In August 2010, cholera was reported. In the Carlisle floods of 2005, there were no reported illnesses because safe drinking water was supplied. However, there was an increase in stress-related illness.

Effects of floods on:

- **people** – death; disease; damage to property; insurance claims; disruption to transport
- **the environment** – landslides; loss of wildlife habitats; crops and animals lost; soil erosion; contamination of water supplies.

Now try this

1. Give a named example and date of flooding in a richer area of the world. **(2 marks)**

2. Explain why flooding in poorer parts of the world is often accompanied by disease and illness **(2 marks)**

Hard and soft engineering

Hard and soft engineering both give ways to manage flood risks, but each has costs as well as benefits.

Hard engineering involves building structures as a defence against flooding.

Dams and reservoirs

River straightening

Building levées, dredging bed

👍 Water stored until needed.

👎 Very expensive and dams trap sediment.

👍 Water moves out of area more quickly.

👎 Increases flood risk downstream.

👍 Relatively cheap; levées only need building once.

👎 Levées block views of the river; dredging needs to be done every year.

Soft engineering means adapting to floods, and using natural processes to help deal with heavy rainfall.

Flood warnings

Preparation

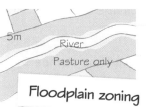
Floodplain zoning

Do nothing

👍 Gives people time to prepare.

👎 Doesn't prevent flooding.

👍 Reduces impact of flooding.

👎 Doesn't always work; expensive.

👍 Buildings on higher land, so they don't flood.

👎 Restricts growth of towns.

👍 Saves money.

👎 Very unpopular – people want protection.

Worked example

tier F&H

Describe the benefits and costs of using a dam to control river flooding. **(6 marks)**

A dam provides a lot of control over a river's discharge because only as much water as is needed can be released. At the same time, a dam can create a reservoir, to provide a reliable water supply, and can be used to generate electricity through HEP. However, dams are very expensive to build. The river water slows down behind the dam and drops its sediment, which can silt up a reservoir, making it less effective. The river also doesn't get to drop its sediment downstream, which may make farmland less fertile.

Now try this

tier F&H

River straightening speeds up a river's flow. Explain why this helps reduce flooding where the river is straightened. **(2 marks)**

EXAM ALERT!

Some students only wrote about the costs of using a dam and not the benefits as well.

Students have struggled with exam questions similar to this – **be prepared!**

Managing water supply

As the population of the UK increases, the demand for water increases too. The areas with the most rain are not the areas with the most population, so supply must be managed.

Water usage in the UK

- Farming and industry use a lot of water, but most is used as a coolant in the power industry.

- Rivers and pipelines are used to **transfer** water from major reservoirs in wet areas to big cities. However, pipelines are very expensive.

- Areas in the south and east have smaller reservoirs. They also rely on groundwater supplies. Drier areas and highly populated areas can suffer **water stress** if rainfall is too low.

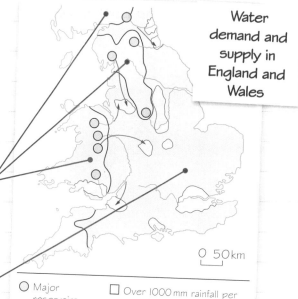

Water demand and supply in England and Wales

○ Major reservoirs
— Rivers
→ Transfers of water by pipeline

☐ Over 1000 mm rainfall per year (high water supply)
☐ Less than 1000 mm rainfall per year (low water supply)
☐ Major urban areas (high water demand)

O 50 km

Areas of **surplus**: areas with lots of rain and low populations.
Areas of **deficit**: drier, highly populated areas.

Case study

You need to know a case study of a dam or reservoir – the economic, social and environmental issues that result and the need for sustainable supplies.

The Elan Valley reservoirs in Powys

When was it built?
Work began in 1893.

What is it for?
Supplies Birmingham with clean, safe drinking water.

Economic issues – creates jobs in Wales; generates electricity, but costs a lot to build and maintain.

Environmental issues – valleys were flooded to create the reservoirs, but now many wildlife species depend on the reservoir.

Social issues – a popular place for recreation and school visits; reduced spread of water-borne diseases in Victorian Birmingham.

Worked example

tier **Fn**

The provision of the future water supplies in the UK has to be sustainable. Suggest **one** or **more** ways that this may be achieved. **(2 marks)**

New houses and industrial sites should be built with facilities for storing rainwater and for reusing wastewater.

EXAM ALERT!

Some students wrote a definition of sustainable water supplies, which the question did **not** require.

Students have struggled with exam questions similar to this – **be prepared!**

Now try this

tier **H**

Explain the problems with using pipelines and rivers to transfer large amounts of water from areas of surplus to areas of deficit. **(4 marks)**

Changes in ice cover

The amount of ice on the Earth's surface keeps changing over time.

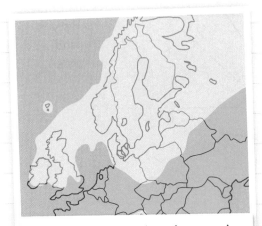

During the last Ice Age, only 18000 years ago, thick ice sheets covered most of the UK. What wasn't covered with ice was frozen, forming permafrost.

Glacials and interglacials

The last Ice Age was part of a period of Earth's history called the **Pleistocene**, which lasted for around 2 million years. Global temperatures fluctuated greatly and glacial periods were interspersed with interglacials – warmer periods.

Holocene

Now we are in a warmer period called the **Holocene**. Much of the ice has melted and covers only around 10% of the Earth's surface – most of that is in Antarctica.

- Antarctica has approximately 14 million km² of ice.
- Greenland has 1.7 million km² of ice.
- Ice caps and glaciers are found in mountainous areas like the Alps, Rockies and Himalayas.

An **ice cap** is like a small ice sheet: less than 50000 km². This one is in the Canadian Arctic.

Glaciers are 'rivers of ice'; they spread out from ice caps down through valleys. This is the Aletsch glacier in Switzerland.

Evidence of change

Here are three ways we know about changes in temperature thousands of years ago:

1. Core samples taken from *deep ice sheets* give a record of changing temperatures from the different chemicals trapped in the layers of ice.

2. Landforms show where there was glaciation.

3. Fossils show where animals lived that we know only tolerate warmer or colder temperatures (e.g. hippo fossils in Leeds, showing it went through a warmer period).

Worked example

tier F n

Complete the following table to show whether the statements in it are true or false. Place a tick in either column to show your answer. **(3 marks)**

Statement	True	False
Glaciers are found only in Antarctica and Greenland.		✓
There are glaciers in Africa.	✓	
There are glaciers in the UK.		✓

Do not include **more** than three answers.

Now try this

Give **two** types of evidence that would show that most of the UK was once covered in ice.

(2 marks)

tier F n

The glacial budget

The glacial budget is the difference between accumulation (snowfall) and ablation (snow melt). A negative budget has more ablation than accumulation and the glacier retreats.

Case study

You need to know a case study of retreating glacier. You need to know why it is retreating and the evidence for it.

- The glacial budget changes between seasons: a lot of accumulation with little ablation in winter, a lot of ablation with little accumulation in summer.

- The glacial budget also varies over time. Since 1950, glacial budgets have been negative.

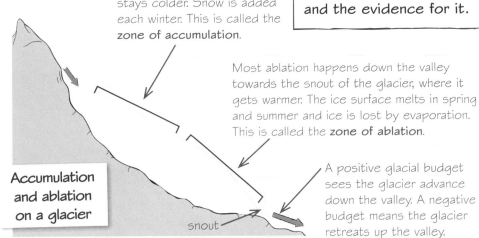

Most accumulation occurs higher up the glacier, where it stays colder. Snow is added each winter. This is called the zone of accumulation.

Most ablation happens down the valley towards the snout of the glacier, where it gets warmer. The ice surface melts in spring and summer and ice is lost by evaporation. This is called the **zone of ablation**.

Accumulation and ablation on a glacier

snout

A positive glacial budget sees the glacier advance down the valley. A negative budget means the glacier retreats up the valley.

Glacial retreat

Warming temperatures are causing glacial retreat. There is lots of evidence to prove this.

debris left by the glacier as it retreats

increases in lakes and rivers fed by meltwater

old paintings and old photos showing where a glacier used to extend to

old maps marking former glacier extent

Evidence of glacier retreat

old measurements made by geographers in the past

recent aerial photos and satellite images

data from glacier measuring devices

Worked example

Explain why glaciers have been retreating since the 1950s. **(6 marks)**

tier **F&H**

For glaciers to be retreating for such a long period, factors must have been making glacial budgets negative year after year. These factors are lower snowfall in winter, reducing accumulation, and higher temperatures making summer melting more significant than previously. Both these factors are associated with global warming. There is common agreement that global temperatures have increased by around 0.8°C in the last 100 years. In landlocked areas such as Switzerland, where there is no cooling effect from the sea, a temperature rise of 1.8°C has been recorded since 1937.

Now try this

Use the templates at the back of this book to help you detail the evidence for glacier retreat and information on causes of glacier retreat for your case study.

tier **F&H**

Glacial weathering, erosion, transportation and deposition

Ice is involved in weathering, erosion, transportation and deposition. Glaciers are extremely effective at eroding and transporting material.

Freeze–thaw weathering

Freeze–thaw weathering is very active in glaciated areas

Water fills a crack or joint in the rock.

Water freezes and the crack is widened.

Repeated freeze–thaw action increases the size of the crack until the block of rock breaks off.

Loose blocks of rock are called scree.

Glacial erosion

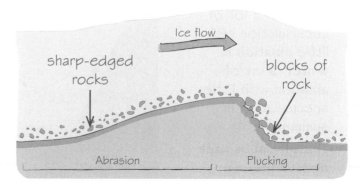

Small sharp-edged rocks and rock particles embedded in the bottom of the glacier wear away and polish the bedrock the glacier is passing over. This is called **abrasion**. Larger rock fragments scrape the bedrock causing **striations**.

Blocks of bedrock freeze to the bottom of the glacier and get plucked out as the glacier moves down the valley. This is called **plucking**.

Transportation

Glaciers are very powerful and can erode and then transport a huge amount of material.

All the material transported by a glacier is called **moraine**. Some is on top of the glacier, some within it and some underneath it.

Rotational slip leads to the formation of a corrie, a circular rock hollow.

See p. 45

For more on rotational slip.

Bulldozing happens when a glacier that has retreated then moves forward again, pushing debris ahead of it.

Deposition

Most deposition happens when the ice melts. Since ice melts most at the **snout** (the furthest point of the glacier), this is where most deposition happens. Deposition also happens if the ice thins or goes around an obstacle making it less able to carry as much load.

Worked example

tier H

What is meant by the term 'rotational slip'? **(1 mark)**

As a corrie forms, ice gets trapped in a hollow and cannot move downhill. But gravity still makes it move in the hollow and this circular motion is called rotational slip.

Now try this

Explain why abrasion sometimes produces smoothing of the bedrock under the glacier, and sometimes scratches (striations). **(4 marks)**

tier H

Glacial erosion landforms 1

Distinctive landforms result from different processes. Glacial erosion produces corries, arêtes, and pyramidal peaks.

A corrie

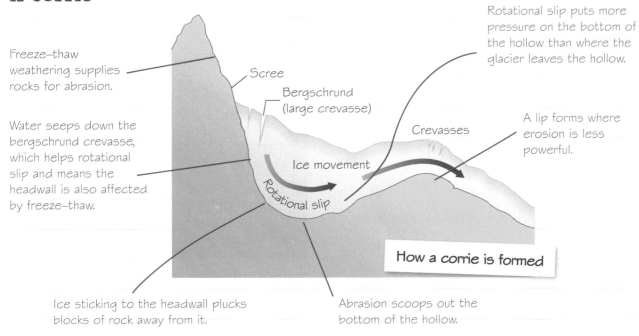

Freeze–thaw weathering supplies rocks for abrasion.

Water seeps down the bergschrund crevasse, which helps rotational slip and means the headwall is also affected by freeze–thaw.

Scree

Bergschrund (large crevasse)

Ice movement

Rotational slip

Crevasses

Rotational slip puts more pressure on the bottom of the hollow than where the glacier leaves the hollow.

A lip forms where erosion is less powerful.

How a corrie is formed

Ice sticking to the headwall plucks blocks of rock away from it.

Abrasion scoops out the bottom of the hollow.

An arête

arête

An arête forms when two corries form back to back

A pyramidal peak

pyramidal peak

A pyramidal peak forms when three corries cut backwards

Worked example

Using the map extract opposite, label a corrie, an arête and a pyramidal peak. **(3 marks)**

This map is from the Ordnance Survey Landranger series, map 115, 1:50 000 scale

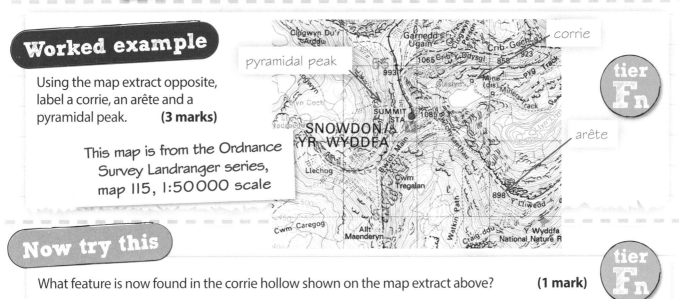

pyramidal peak

corrie

arête

tier F n

Now try this

What feature is now found in the corrie hollow shown on the map extract above? **(1 mark)**

tier F n

Glacial erosion landforms 2

Glacial erosion also produces truncated spurs, glacial troughs, ribbon lakes and hanging valleys.

A valley before glaciation

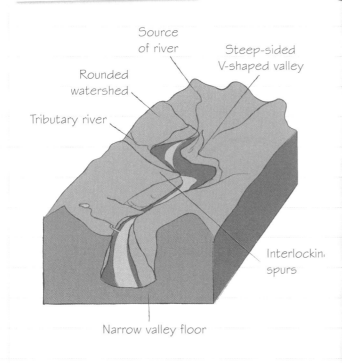

- Source of river
- Steep-sided V-shaped valley
- Rounded watershed
- Tributary river
- Interlocking spurs
- Narrow valley floor

Worked example

tier F n

Which of these glacial erosion landforms does this photo from New Zealand show? Mark your answer with a tick:

Corrie	
Hanging valley	✓
Ribbon lake	

(2 marks)

A valley after glaciation

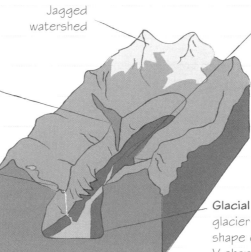

Jagged watershed

Truncated spurs are formed when the glacier pushes straight through the valley and removes the ends of the interlocking spurs of the old river valley.

Tributary in a hanging valley with a waterfall. **Hanging valleys** are old tributary valleys, occupied by small glaciers that can't erode down as far as the main glacier.

Ribbon lakes are long and thin and form when a glacier erodes a long, thin section of the valley floor more deeply than the rest. This usually happens when the ice gets thicker where another glacier joins as a tributary or where a valley has been blocked by a terminal moraine and meltwater builds up behind.

Glacial trough is a glaciated valley. A glacier has deepened and changed the shape of a former river valley from a V shape to a U shape.

Check your diagram carefully to make sure everything is where it should be and all labelling is in the right place, with the head of the arrow actually **touching** the feature identified.

Now try this
tier F&H

Draw and annotate a cross section of a glaciated valley to include the following features:

U-shaped valley   **Hanging valley**   **Flat valley floor**   **Waterfall**   **(4 marks)**

Glacial landforms of transportation and deposition

Glacial transportation and deposition produces drumlins and four types of moraine: lateral, medial, ground and terminal.

Moraines

Lateral moraine

A build up of material along the side of glaciers. When the glacier melts this forms a ridge along the valley sides.

Terminal moraine

Most deposition occurs at the snout where great ridges of material pile up. Terminal moraines mark the furthest extent of the glacier (terminal means end).

Medial moraine

When a tributary glacier joins the main glacier, lateral moraines merge towards the middle of the main glacier. When the glacier melts the moraine is deposited in a ridge down the middle of the valley.

Ground moraine

As the glacier melts it drops material known as boulder clay or till all over the valley floor, leaving hummocky ground.

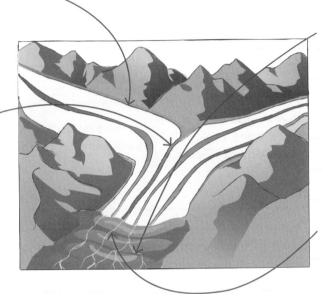

Drumlins

Drumlins are low hills (30–40 metres high) that are egg shaped: blunt at one end, tapered at the other. They show what direction the ice flowed in as the tapered end points in the direction of the flow.

Drumlins form when melting ice is pushed forward over a lowland area, depositing material as it goes. Where there are obstacles, deposition is increased making the blunt end of the drumlin. Further deposits are moulded into the tapered end as the glacier continues its course.

Lots of drumlins can occur together. This is called a drumlin 'swarm'.

Worked example

Identify the depositional landform in the photo above. **(1 mark)**

Drumlins.

Now try this

Study the diagram of moraines in the diagram at the top of this page.
Explain why there are two other ridges behind the terminal moraine. **(4 marks)**

Alpine tourism – attractions and impacts

Alpine areas are popular tourist destinations. You need to know the impacts tourists have on these areas.

Case study

You need to know a case study of tourism in an Alpine area.

Attractions of Alpine areas

- winter snow for skiing, snowboarding
- lots of facilities for winter sports: ski lifts, ski resorts, hotels
- long history of ski tourism (e.g. 250 years for Chamonix)
- easy access to mountains – cog trains, cable cars
- beautiful mountain scenery
- shops, spas, museums
- all-year glaciers for summer visits (e.g. rail trip to Mer de Glace)
- marked hiking trails for summer walking
- rock climbing, canoeing, pony trekking in summer

Impacts of tourism

Positive	Negative
Economic Creates jobs and boosts the local economy.	Social Young people get jobs in tourism rather than farming: difficult to keep farms going.
Opportunities to earn a living from tourism all year round, not just in winter.	Shops and services are focused on tourists, not locals' needs.
Social Young people from all over the world come to work.	Traffic congestion in villages and along narrow mountain roads.
Jobs for young people who might otherwise have left the rural area for city life.	Environmental Clearing forests to build resorts and ski slopes (clearing trees also increases avalanche risks).
Environmental Tourism brings in money that can be used to protect the natural environment.	The fragile Alpine environment is easily damaged by walkers, skiers and vehicles.

Worked example

tier F&H

Give one or more examples of conflicts that can occur over the use of Alpine tourist areas.   **(4 marks)**

⬆ You also need to know about management strategies and their success in your case study area. These are outlined on page 49.

Tourists who come to enjoy peaceful walks in the mountain scenery may object to mountain bikers using the same mountain trails. Some parties of young skiers enjoy noisy après-ski parties which older skiers might resent. Farmers who need gates to stay closed to protect livestock have to deal with walkers who forget to close gates.

Now try this

Use the template on page 114 to complete a revision flash card for your case study. Make sure you have details on the attractions for tourists in your case study area and the economic, social and environmental impacts that tourism has brought.

Management of tourism and the impact of glacial retreat

Alpine areas need to be managed to reduce the impact of tourists on the fragile environment.

Make sure you have case study details to add to generic points about tourism management strategies

Case study

Your case study of tourism in an Alpine area should include how the area is managed.

Managing the impacts of tourism

- environmentally friendly public transport systems to reduce congestion
- sustainable energy sources used where possible
- car access to towns and resorts restricted
- the development of new ski stations strictly controlled
- local community supported by creating all year round jobs and funding traditional skills
- conservation areas set up to protect plants and animals
- footpaths maintained by repairing erosion
- avalanche risk monitored and tourists kept away from high-risk areas
- damaged areas of the ski slopes fenced off and bare patches reseeded

Impacts of glacial retreat

Global warming not only leads to glacier retreat, it also makes snowfall less reliable. So tourism can be hit for both winter sports and glacier viewing in summer.

- Economic – fewer tourists would mean fewer jobs and less money for the local economy.
- Social – fewer jobs means that people will move away from the mountain regions. There will be fewer services for the remaining local people.
- Environmental – global warming and glacial retreat could mean an increase in natural hazards, including avalanches and flooding.

Avalanche hazard

- Avalanches are sudden downhill movements of snow and ice (and often rocks).
- The avalanche descends at great speed, averaging 40–60 km/h.
- Avalanches are very hard to predict.

Worked example

What is meant by the term 'fragile environment'?
(1 mark)

A fragile environment is one where plants and animal species are very delicately balanced due to harsh conditions. Damage to the environment takes a very long time to be repaired.

Now try this

Many Alpine communities depend on glacial meltwater for their water supply, to water crops and often for energy from hydroelectric power (HEP). Describe how glacial retreat might affect this use. **(2 marks)**

Waves and coastal erosion

Destructive waves are the main cause of erosion in a coastal zone.

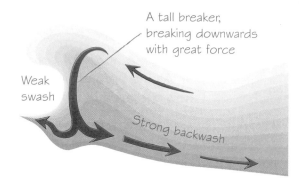

A tall breaker, breaking downwards with great force

Weak swash

Strong backwash

In a **destructive wave** the swash is weak and the backwash is strong which means material is dragged back down a beach into the sea.

Processes of coastal erosion

- **Hydraulic power:** the sheer weight and impact of water against the coastline, particularly during a storm, will erode the coast. Also waves compress air in cracks in the rock, forcing them apart and weakening the rock.
- **Abrasion:** breaking waves throw sand and pebbles (or boulders) against the coast during storms.
- **Attrition:** the rocks and pebbles carried by the waves rub together and break down into smaller pieces.
- **Solution:** chemical action on rocks by seawater dissolve some rocks, especially limestone.

Weathering

Mechanical, chemical and biological weathering all happen at the coast.

See p. 11

For more on weathering.

Mass movement

Mass movement is the downhill movement of material under the influence of gravity. The different types of mass movement depend on:

- the material involved
- the amount of water in the material
- the nature of the movement e.g. falls, slips or rotational slide.

Worked example

tier **F n**

Complete the gaps in this sentence using the correct selections from the following words:
deposited, ~~erosion~~, movement, transported, ~~weathering~~.　　　　(3 marks)

The material produced by mass _movement_ and weathering is usually eroded and _transported_ by waves and _deposited_ elsewhere along the coast.

Remember, in a cloze exercise like this there will be some incorrect answers. Try to identify these and cross them out.

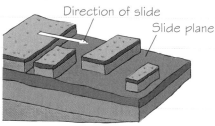

Direction of slide

Slide plane

Loose, wet rocks slump down under the pull of gravity along curved slip planes

Sliding happens when loosened rocks and soil suddenly tumble down the slope. Blocks of material might all slide together.

Slumps happen when the rock (often clay) is saturated with water and slides down a curved slip plane.

Now try this

This answer should cover three types of weathering process.

Describe the weathering processes found in the coastal zone.　　　　(6 marks)

tier **H**

Coastal transportation and deposition

Waves do most of the transportation along the coast. They deposit material when they lose the energy to carry it further.

Longshore drift

Waves often approach a coastline at an angle under the influence of the prevailing winds, but sand and pebbles roll straight back down the beach under the influence of gravity. In this way sand and pebbles are moved along a beach.

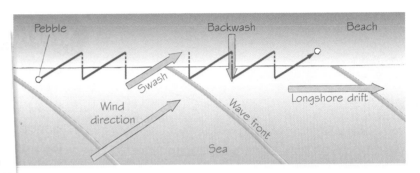

Pebble · Backwash · Beach · Swash · Wind direction · Wave front · Longshore drift · Sea

Transportation

Waves **transport** material by:

Traction – large boulders are rolled along the sea bed by waves

Saltation – smaller stones are bounced along the sea bed

Suspension – sand and small particles are carried along in the flow

Solution – some minerals are dissolved in seawater and carried along in the flow

Deposition

A constructive wave

Crest breaks forwards

The wave is long in relation to its height

Strong swash

Weak backwash

The load carried by waves is deposited by constructive waves. Different factors influence deposition, for example:
- sheltered spots (e.g. bays)
- calm conditions
- gentle gradient offshore causing friction.

All reduce the wave's energy.

Worked example

You could use annotated diagrams to show the main features of the waves.

 tier F&H

Describe the differences between a constructive wave and a destructive wave.   **(4 marks)**

Constructive waves deposit material on beaches because they break gently on the beach and their strong swash carries material up the beach, while their weak backwash does not erode the material already on the beach. Destructive waves erode material from beaches because the backwash of these waves is much stronger than their swash and this drags material back down the beach into the sea.

Now try this

Naming just means give the names. You don't need to add any description.

 tier Fn

1 Name the four processes by which waves transport sediment.   **(4 marks)**
2 Is a beach a feature of deposition or erosion?   **(2 marks)**

 tier F&H

Landforms of coastal erosion

Landforms resulting from erosion include headlands and bays, cliffs and wave-cut platforms, caves, arches and stacks.

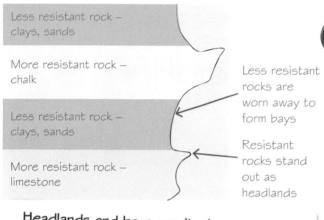

Less resistant rock – clays, sands

More resistant rock – chalk

Less resistant rock – clays, sands

More resistant rock – limestone

Less resistant rocks are worn away to form bays

Resistant rocks stand out as headlands

Headlands and bays result when a coastline has bands of more and less resistant rocks.

Worked example

tier F n

Explain why beaches are often found in bays. **(2 marks)**

Bays are more sheltered and there is a gentle gradient offshore, so waves have less energy and are more likely to deposit their load. Also when bays are curved, waves are refracted (bent in a curve) and this also slows them down.

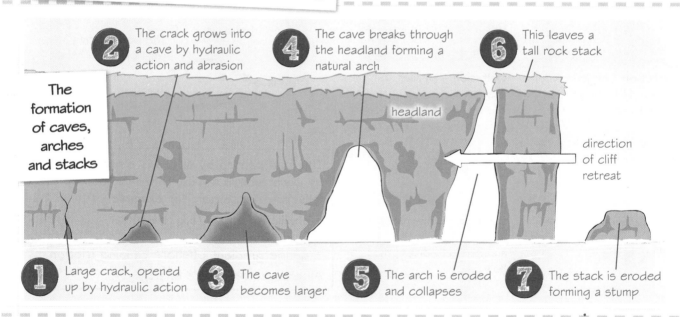

The formation of caves, arches and stacks

2 The crack grows into a cave by hydraulic action and abrasion

4 The cave breaks through the headland forming a natural arch

6 This leaves a tall rock stack

headland

direction of cliff retreat

1 Large crack, opened up by hydraulic action

3 The cave becomes larger

5 The arch is eroded and collapses

7 The stack is eroded forming a stump

Cliff retreat and the formation of a wave-cut platform

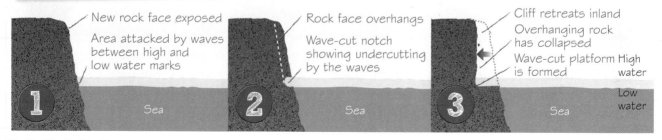

New rock face exposed

Area attacked by waves between high and low water marks

Rock face overhangs

Wave-cut notch showing undercutting by the waves

Cliff retreats inland

Overhanging rock has collapsed

Wave-cut platform is formed

High water

Low water

1 Sea   **2** Sea   **3** Sea

Now try this

Draw and annotate a diagram to show the formation of a wave-cut platform. **(4 marks)**

tier F&H

Landforms resulting from deposition

Landforms resulting from deposition include beaches, spits and bars.

Beaches

Beaches are accumulations of sand and shingle formed by deposition and shaped by erosion, transportation and deposition.

Beaches can be straight or curved. Curved beaches are formed by waves refracting, or bending, as they enter a bay.

Beaches can be sandy or pebbly (shingle). Shingle beaches are usually found where cliffs are being eroded and where waves are powerful. Ridges in a beach parallel to the sea are called berms and the one highest up the beach shows where the highest tide reaches.

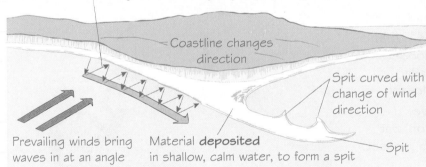

Material moved along beach in a zig-zag way by longshore drift

Coastline changes direction

Spit curved with change of wind direction

Prevailing winds bring waves in at an angle

Material **deposited** in shallow, calm water, to form a spit

Spit

Spits are narrow beaches of sand or shingle that are attached to the land at one end. They extend across a bay or estuary or where the coastline changes direction. They are formed by longshore drift powered by a strongly prevailing wind.

Worked example

 tier F&H

Explain the formation of a bar as shown in the photograph. **(4 marks)**

A bar forms in the same way as a spit, with longshore drift depositing material away from the coast until a long ridge is built up. But unlike a spit, a bar then grows all the way across a bay so that a stretch of water is cut off and dammed to form a lagoon.

Loe Bar in Cornwall

Now try this

Give **two** sources for the material that beaches are made from. **(2 marks)**

 tier Fn

Rising sea levels

Some coastal areas are affected by rising sea levels.

Reasons for rising sea levels

- Global warming is thought to be the reason for rising sea levels. Increases of up to 3 mm per year have been recorded for several decades.
- Global warming experts warn of sea level rises in the region of 28–43 cm by 2100.
- Sea level rises are mostly due to thermal expansion of the water as global temperatures increase – warm water takes up more room.

Case study

You need to know a case study to illustrate the economic, social, environmental and political impact of coastal flooding. This can be anywhere in the world, not just the UK.

- Sea level rises are not the same everywhere. For example, the land in the north of Britain was pushed down by the weight of the ice during the last Ice Age and it is still rising back up again now the weight has gone.
- Some areas are more at risk from sea level rises than others, for example, low-lying coastal land.

Case study

Impacts from rising sea levels

Location: UK

Economic Loss of agricultural land would be a big problem, particularly in East Anglia.

Economic What if London was flooded? The UK economy could be very badly affected and economic losses colossal.

Political Some settlements will be defended from flooding and erosion, but there will not be enough money to protect everyone at risk.

Social Storm surges (major coastal floods) are very dangerous. 307 people were killed by a storm surge along the east coast of the UK in 1953.

Environmental Destruction of natural ecosystems; e.g., the Essex marshes.

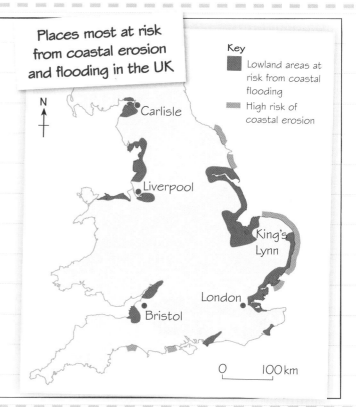

Places most at risk from coastal erosion and flooding in the UK

Key

■ Lowland areas at risk from coastal flooding

■ High risk of coastal erosion

Carlisle, Liverpool, King's Lynn, London, Bristol

0 100 km

Worked example

tier F n

State two threats to a coastal area from rising sea levels. **(2 marks)**

Storm surges; loss of agricultural land to flooding.

Now try this

tier F&H

Use the case study revision template on page 114 to revise the details you need to know about the economic, social, political and environmental consequences of coastal flooding in your case study.

Detailed knowledge will be very valuable in your exam, but make sure you only use details that are **relevant** to the question you are asked.

Coastal management

Coasts can be managed in different ways to defend them against the sea.

Hard engineering

A traditional approach to protecting coasts involves building structures.

Case study

You need to know a case study of coastal management that looks at the costs and benefits of the strategies that have been used. See also page 56.

Sea wall

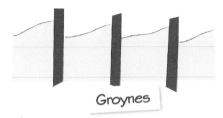
Groynes

Rock armour

- 👍 Protects cliffs and buildings
- 👎 Expensive

- 👍 Stops sea removing sand
- 👎 Exposes other areas of coastline

- 👍 Rocks absorb wave energy
- 👎 Expensive

Soft engineering

Working with nature to help maintain the present coastline.

Beach nourishment

Managed retreat

Dune regeneration

Marsh creation

- 👍 Sand reduces wave energy and maintains tourism
- 👎 Expensive

- 👍 People and activities move inland
- 👎 Unpopular with local residents

- 👍 Relatively cheap
- 👎 People resent having to stay out of replanted areas

- 👍 Creates wildlife habitat
- 👎 Farmland destroyed and farmers have to be compensated

Worked example

Part (b) says describe. You need to **develop** your answer to get the marks.

tier **Fn**

This photo shows coastal management at Hornsea in Yorkshire:

(a) Identify the coastal management feature at A.　**(1 mark)**

Groyne

(b) Briefly describe how coastal management feature B protects the coastline.　**(2 marks)**

Rocks prevent damage to the sea wall by absorbing wave power and trapping beach sand.

Now try this

tier **H**

Outline the main advantages and disadvantages of hard engineering methods used to protect coasts. **(6 marks)**

Cliff collapse

Coastal erosion can lead to cliff collapse. There are many different factors that contribute to cliff collapse.

Your cliff collapse case study needs to include details on:
- rates of coastal erosion
- reasons why some areas are susceptible to undercutting by the sea and collapse
- how people may worsen the situation and the impact on people's lives and the environment.

Case study

You need to know a case study of an area that has recently had a cliff collapse or which is threatened by cliff collapse.

Factors that contribute to cliff collapse

- coastal erosion
- weathering
- arrangement of rock layers, e.g. permeable sand overlying impermeable clay causing landslides because of saturation at the base of the sand layer
- soft, weak rock types (e.g. clays)
- heavy rain
- strong winds blowing over a long stretch of water (fetch)
- human influences, including building near cliff edges (weight), changing drainage patterns that increase saturation

Impacts of cliff collapse

- very dangerous to people on the cliffs or on the beach below
- houses and businesses (e.g. caravan parks, golf courses) may be damaged
- roads, railways and paths may be destroyed
- houses and businesses near eroding cliffs are very hard to sell or insure
- wildlife habitats destroyed
- increased deposition further along coast

Worked example

tier **F** n

Give one example of how cliff collapse can threaten the environment.   **(2 marks)**

Seabirds such as guillemots and kittiwakes nest on cliffs and a cliff collapse during the nesting season would have a very significant impact on these birds.

EXAM ALERT!

Some students struggled to answer this question because they had only revised impacts of cliff collapse on people's lives.

Students have struggled with exam questions similar to this – **be prepared!**

Now try this

Give an example of a coastal area that is experiencing rapid coastal erosion and cliff collapse. Indicate the rate of coastal retreat.   **(2 marks)**

tier **F** n

Managing the coast

Different strategies are used to manage the coastline. These have costs and benefits.

You need to know a case study of coastal management and the costs and benefits of different strategies.

Case study

Coastal management

Location: Mappleton on the Holderness coast in East Yorkshire.

What is the rate of erosion?

Upwards of 1.8 m a year.

Reasons for rapid erosion

- Easily eroded boulder clay is also prone to slumping when saturated.
- Exposed to strong waves (fetch).
- Further up the coast, groynes have been built to stop the movement of sand southwards by longshore drift, and so the natural protection given by a beach has been removed.

Impacts on people's lives

Property prices have slumped for houses near the cliff.

Management strategies

Two rock groynes and rock armour.

Benefits of management strategies:
Village of 100 people protected and the B1242 road link along coast protected.

Costs

- £2 million for 100 people.
- Lost farmland and property.

Successful?

The scheme is a success for Mappleton, but what are the impacts elsewhere?

Management methods

- Many different methods can be used.
- They can be changed over time.
- They work best when they are used in combination.

Worked example

tier **F̶ H**

People directly affected by coastal erosion often favour hard engineering. Give an example of a group of people likely to oppose this view and explain why they have this opinion.   **(2 marks)**

Local people who do not live in the threatened area might object to having to pay extra taxes to fund expensive coastal defence schemes.

EXAM ALERT!

Some students wrote about the costs and benefits of hard engineering rather than thinking about what the question was asking for.

Students have struggled with exam questions similar to this – **be prepared!**

Now try this

tier **H**

Explain the management strategies used to control coastal erosion in a named example of a coastline you have studied.   **(6 marks)**

Make sure you fully explain the strategies, rather than just describing them. A good answer would refer to more than one strategy because the question asks for strategies, not just strategy.

Coastal habitats

Coastal areas provide different habitats and environments.

Salt marshes

Salt marshes form in sheltered conditions at the coast either in estuaries or bays. They provide valuable habitats for birds and other marine life and plants. They also act as buffer zones against floods

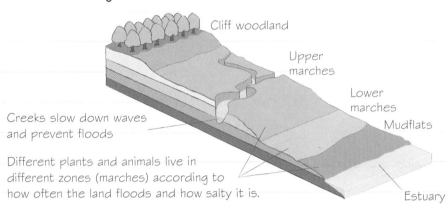

Cliff woodland

Upper marches

Lower marches

Mudflats

Creeks slow down waves and prevent floods

Different plants and animals live in different zones (marches) according to how often the land floods and how salty it is.

Estuary

Conservation of salt marshes – managed retreat

Realising how important salt marshes are for wildlife and for flood protection, the sea, in places has been allowed to break through defences and flood large areas to create marsh habitat. This is called **managed retreat**.

Salt marsh in Norfolk showing the creeks that characterise this habitat.

Salt marshes under threat from sea level rises and development (draining for farming)

↓

Managed retreat strategies encourage salt marshes to develop

↓

Wildlife conservationists like RSPB are interested in salt marsh conservation

↓ ↓ ↓

| More managed retreat on RSPB reserves | Control over visitor numbers and where visitors can go | Encourage education of visitors: responsible tourism |

tier F&H

Worked example

Describe one strategy that makes sure a coastal habitat is conserved in a sustainable way.   **(4 marks)**

Using boardwalks means that visitors can walk through a delicate habitat without damaging the vegetation cover and without straying into areas that need complete protection. This is sustainable because it means future generations will be able to use the habitat in the same way.

Now try this

Name **three** issues that threaten the coastal habitat you have done for your case study.

tier F n

Population explosion

The world's population has grown exponentially.

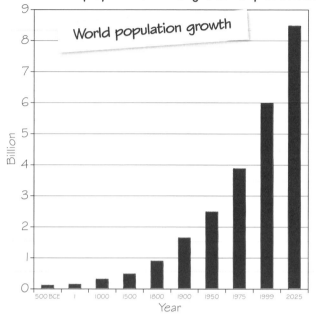

World population growth

- Population growth was slow until the start of the 19th century.
- The world's population took thousands of years to get to 1 billion (in 1804).
- It took a 123 years to get to 2 billion.
- It reached 3 billion in another 33 years.
- By the end of the 20th century, this had doubled to 6 billion.
- It only took 11 more years to get to 7 billion.
- Experts think global population will stabilise at 9.5 billion by 2050.

Natural increase

Population grows when the birth rate is higher than the death rate.
This is called **natural increase**.

Medical progress lowers the death rate:	These Social, Economic and Political factors keep the birth rate high:
• Primary healthcare – preventing disease (e.g. immunisation) • Secondary healthcare – treatment of illnesses	• Family planning is not widely used • Women receive little education and marry young • Families of five or more children are considered normal • Children work to help support the family • Parents rely on their children in old age • Governments do not provide family planning help • Religions may not approve of birth control

Birth rates vary

Birth rates tend to be high in poorer countries, whereas they become lower with economic development. For example:

- Botswana's birth rate: 31 births per 1000 people.
- UK's birth rate: 11 births per 1000 people.

Death rates tend to be low in most countries (rich and poor) making natural increase high in poorer countries, but low or even negative in richer countries.

Worked example

tier H

What is meant by an 'exponential rate' of population growth?   **(1 mark)**

This is when population growth becomes more and more rapid as time goes on.

Now try this

What is meant by the term 'life expectancy'?

'What is meant by' questions are looking for a **definition**.

(1 mark)

tier Fn

Demographic transition

Countries pass through different stages of population growth, as shown in the five stages of the Demographic Transition Model.

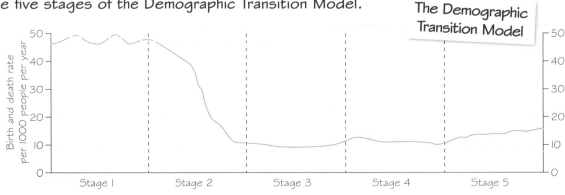

The Demographic Transition Model

Stage	1	2	3	4	5
Birth rate	High	High	Decreasing	Low	Low
Death rate	High	Decreasing	Low	Low	Low
Natural increase	Low	Becoming high	High becoming low	Low	Natural decrease
Countries	None – all countries have progressed into another stage.	Poor countries with low levels of economic development, many of them in Africa.	Poor countries with improving levels of economic development, many in south and east Asia, Mexico, Brazil.	Rich countries in North America, Australasia, Japan and many European countries.	A few rich countries with very low birth rates, mainly in Europe.

Authorities provide family planning clinics

Small families accepted as normal

Uncertainty about future economic growth

More people living in cities: less need for big families

With increasing wealth, children contribute less to family income

Factors leading to fall in birth rate

Agriculture becomes more mechanised – less need for big families to farm the land

Female income vital to family budget

National population policy

Women more career orientated

Equal female education

Free government contraceptive supply

Worked example

tier Fn

Study the diagram above. Explain why increasing wealth often means a declining birth rate.

(2 marks)

High birth rates are associated with poverty because more children means more workers to support the family. So as wealth increases, families can feed themselves just from what the parents earn.

Now try this

tier F&H

Explain why the third stage of the Demographic Transition Model is sometimes called the Late Expanding stage.

(4 marks)

Explain questions want you to give reasons for something.

Population structure

The population structure of a country refers to how many people of each sex are found in different age groups. Population structure is shown in population pyramids.

Population pyramids

Changes in population structure are related to stages in the Demographical Transition Model.

Stage 1

Very wide base – very high birth rate.

Jagged sides because population frequently affected by natural disasters, e.g. crop failure, disease.

Narrow top – death rates high for all ages.

Stage 2

Base a bit narrower – birth rate declines slightly.

Sides more regular as food supply becomes more dependable – reliable agriculture.

Top still very narrow – very few survive to old age.

Stage 3

Narrowing base – decrease in birth rate.

Straighter sides – more people live to older ages.

Stage 4

Straight sides – steady low birth rate.

Concave base – birth rate starting to decline.

Wide apex – high life expectancy.

Stage 5

Bulge in middle – more middle aged than young people.

Broader at top – more and more old people.

Narrow at base – falling birth rate.

This question is only worth 2 marks so make sure you only consider the dependancy ratio. You don't need to write about stages 4 and 5 of the DTM.

tier **F&H**

Stages 4 and 5 are marked by an increasing dependency ratio. What does the dependency ratio measure? **(2 marks)**

The dependency ratio is the ratio between the dependent population (children and retired old people) and the independent population (the working part of the population).

Which of these features of a population pyramid would indicate a high birth rate? **(1 mark)**

     a narrow base and a bulging middle   **a narrow top**   **a broad base**

tier **Fn**

Growing pains

A sustainable population has a growth rate that does not threaten the survival of future generations. In many parts of the world, the rapid growth rate is not sustainable.

Problems of rapid population growth in developing countries

In rural areas:

- overgrazing and over-cultivation
- water, land and air pollution
- deforestation, land degradation, soil erosion and desertification
- shortages of clean water supplies
- lack of basic public services (sanitation, electricity, schools, clinics).

In urban areas:

- overcrowding and the growth of shantytowns
- water, land and air pollution
- traffic congestion
- inadequate public services.

In the country as a whole:

- shortages of resources, food and raw materials
- unemployment and under-employment
- lack of money for basic healthcare and schooling
- low living standards and rising crime
- economy burdened by huge international debts
- unstable governments and political coups
- tension between different tribes and races.

Unsustainable population growth

This population pyramid of Zimbabwe in 2010 shows the results of rapid population growth. Half the population is under 25 years old. It may be difficult for all these people to find jobs and homes. It may be difficult to build the schools, hospitals and infrastructure to cope with them all.

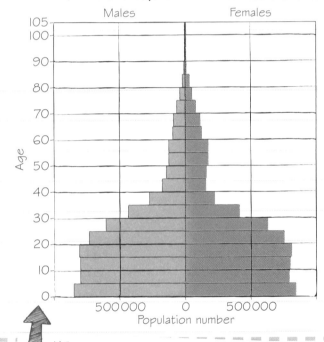

When you are working with graphs, use a ruler to help you read off the values correctly. If you need to draw or complete a bar graph for an exam question, make sure you use a ruler to draw a straight line

Worked example tier Fn

Study the table above. Identify one social, one economic and one political implication of rapid population growth. **(3 marks)**

Rising crime is a social implication, unemployment is an economic implication and unstable government is a political implication.

EXAM ALERT!

Make sure you know the **difference** between social implications and political implications.

Students have struggled with exam questions similar to this – **be prepared!**

'Explain why' questions are asking you to give **reasons** for something.

Now try this tier F&H

Explain why the education of women is a factor in reducing the rate of population growth. **(4 marks)**

Managing population growth

Some countries have adopted policies to control rapid population growth, you need to know how effective these policies are.

Case study

You need to know two case studies: one on China's policy since the 1990s and one of a non-birth control population policy.

Case study

China's population policy

Strategy used

- In 1979 China began the one-child policy.
- Couples who had only one child received economic rewards and welfare benefits, e.g. better housing and free education.
- Couples who had more than one child were fined and did not get any of the benefits. There are now exceptions for rural areas and ethnic minorities.
- Some women were forced to have abortions very late into pregnancy.

Positive results

- China's population is around 300 million lower than it would have been without the one-child policy: the policy worked.

Drawbacks and problems

- Baby girls are not valued as much as baby boys. Many baby girls have been abandoned in orphanages.
- The preference for boys means the Chinese population has more men than women, causing social problems.
- Single children will bear the brunt of an ageing population: each couple could have four parents and eight grandparents to support.
- There may be a shortage of workers in the future to maintain China's massive pace of industrialisation.

Case study

Indonesia's population policy

Indonesia has one of the largest populations of any country and it is growing fast. Indonesia is made up of thousands of islands, but most Indonesians (nearly half) live on just one of them – Java. Java became overcrowded, with many social problems.

Strategy used

Transmigration policy – since the 1960s, millions of people have been moved from islands with high population densities to islands with low densities, like Papua or Sumatra.

Positive results

Around 20 million people have moved.

Drawbacks and problems

- The strategy has not reduced population growth – or even tackled it at all.
- Deforestation increased rapidly when lots of people moved to low-density islands.
- Migrants have not always been welcomed in their new homes.
- Islands such as Java and Bali have rich soils; other islands are harder to farm.

EXAM ALERT!

Don't waste time writing a sentence, the letter is sufficient.

Students have struggled with exam questions similar to this – **be prepared!**

Worked example   tier **Fn**

Is China's one-child policy an example of:

A government having a significant influence on population growth

B government failing to have a significant influence on population growth?   **(1 mark)**

A

Now try this   tier **H**

Compare China's population policy since the 1990s with a country that uses a different approach to controlling population growth.   **(4 marks)**

Ageing population

An ageing population can have an impact on the economic development of the country.

There will be a high proportion of **elderly dependents** (aged 65 and over).

This is a prediction of what the UK's population will look like in 2030

There will be less in the **young dependent** category, creating a high dependency ratio when this group enters the working population.

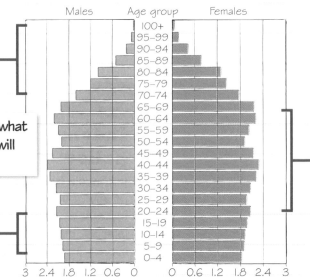

Males   Age group   Females

100+, 95–99, 90–94, 85–89, 80–84, 75–79, 70–74, 65–69, 60–64, 55–59, 50–54, 45–49, 40–44, 35–39, 30–34, 25–29, 20–24, 15–19, 10–14, 5–9, 0–4

3 2.4 1.8 1.2 0.6 0   0 0.6 1.2 1.8 2.4 3
Population number (millions)

Case study

You need to know a case study of the problems and management strategies in one EU country with an ageing population.

The current **working population** (15–64) will create a large elderly dependent group over the next 30 years.

Problems associated with an ageing population

State pensions
Can the state afford to pay pensions when many are living so long after retirement?

Smaller working population
Who is going to work and pay all the tax so the state can afford so many old people?

Increasing health and care needs
How will countries afford to pay for welfare and medical services?

Make sure you revise the problems and strategies for your own case study location.

Worked example

tier **F&H**

Study the population pyramid above. How might a rising birth rate improve the UK's ageing population problem? **(4 marks)**

A rising birth rate would reduce the UK's dependency ratio because it would mean there would eventually be a bigger working population to offset the growing number of elderly dependents. A bigger working population can pay the taxes the state needs to fund care for the elderly and it can also provide the people – care workers and medical staff – to look after elderly people.

Management strategies

Different EU countries have different strategies. Some, such as Germany, Sweden or France, are **pro-natalist**: encouraging people to have more babies. Pro-natalist strategies include:

- free childcare for everyone
- child benefit payments for each child
- paternity leave – Sweden gives fathers 13 months' leave at 80% of their salary.

Now try this

tier **H**

Give a detailed account of how your EU case study attempts to deal with the problems of an ageing population. **(6 marks)**

Remember to outline the problems your case study country is facing too; try to always quote precise details, such as life expectancy and birth rate if you know the

Migration: push and pull

Migration happens as a result of people making decisions based on push and pull factors, which can have positive impacts and negative impacts.

Push factors are the disadvantages of living in a place	→ Migration	**Pull factors** are the advantages of where people want to move to
Harsh climate		Cheap land
Innaccessibility		No hazards
Divorce		Marriage and family ties
Ill health		Plenty of work
Unemployment		High wages
Poverty		Better lifestyle
Shortage of housing		Good welfare services
Civil war		Personal security
Ethnic cleansing		Freedom of speech

Forced migration is compulsory migration. It is due to **push** factors. Forced migrants are **refugees**.

Voluntary migration is when a person makes the decision to move. **Pull** factors are often most important here.

Positive and negative impacts of international migration

Migrants bring new or special skills

Send money back to families where it is spent locally on services

Willing to take jobs not wanted by receiving country's own nationals

Advantages

Earn more money and gain a higher standard of living

Transfer of knowledge, enabling economic development in receiving country

Cultural exchange of ideas and lifestyles

Loss of labour in home country when young people move

Family separation, especially husbands from wives and children

Loss of trained people with skills needed in country of birth

Disadvantages

Strain on resources for host country

Increase in racial tension and discrimination

Obstacles to population movements

In general, modern developments have made travel quicker and easier, and people know more about the rest of the world because of developments in the media and communications. But most countries also put up obstacles to free movement of people. Examples would be visa and entry permits, and border guards and fences.

Worked example
tier F n

Give two reasons why people often want to move from the countryside to the city. **(2 marks)**

In developing countries living in the countryside is often hard because the work is difficult and not very well paid. Life in the city seems much more attractive as there are many more jobs and they might pay more for less hard work.

Now try this
tier F n

What is meant by the term 'economic migrants'?   **(1 mark)**

EU movements

Population movements occur within the EU and into the EU from countries outside it. These movements have both positive impacts and negative impacts.

The enlargement of the EU

- In 2004, the EU's enlargement meant workers from ten low-income new member countries could now live, travel and work in any EU country.

- By 2008, the 27 EU member states were receiving nearly 2 million migrants from each other a year.

- At the same time, 1.8 million immigrants were coming to the EU from non-EU countries per year. Moroccans were the largest group of migrants, followed by people from China, India, Albania and Ukraine.

Impacts of population movement within the EU

Positive impacts	Negative impacts
👍 Migrants helped boost declining populations in richer EU countries.	👎 Because migrants will work for less, some feel they take jobs away from citizens of the host country.
👍 Migrants earned a lot more working in richer countries than they would at home.	👎 Migrants use health and social services, which puts more strain on services which are already stretched.
👍 Migrants are willing to take low-paid, unskilled 'dirty' jobs.	👎 A lot of economic migration from one country can result in skills shortages and a shortage of workers.

Migration from outside the EU

Many of the migrants coming to the EU from the rest of the world are economic migrants and they include people from countries such as Australia or the USA, as well as poorer countries.

People from poorer countries outside the EU sometimes pretend to be asylum seekers because they can't get into the EU any other way. Unfortunately, this has made many in host countries distrust genuine asylum seekers, too.

Worked example

tier F&H

Explain the push and pull factors that might encourage someone to leave the UK and move to another EU country. **(4 marks)**

Push factors from the UK often include the weather. A pull factor would be warmer climate and less rain that countries like Spain have. These factors would apply especially to people who have retired. Another push factor might be the high cost of education in the UK. Students might choose to study in another EU country, for example Sweden, where there are pull factors such as opportunities to learn another language.

Now try this

tier Fn

Name an international event that has increased the number of people seeking asylum in EU countries.

(2 marks)

Urbanisation goes global

Urbanisation is the increase in percentage of people living in urban areas – but why is urbanisation occurring so much faster in poorer countries than in richer countries?

Urban population in the rich and poor world

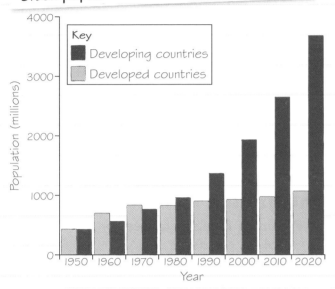

Key
- ■ Developing countries
- ■ Developed countries

(Graph: Population (millions) vs Year, 1950–2020)

- In richer countries, urbanisation can be over 90%. At this level, the rate of urbanisation then slows right down. **Counter-urbanisation** may occur, where people move from the cities to the countryside, looking for better quality of life.
- **Urban growth** is the increase in size of cities and / or their population.
- When cities reach 10 million population, they are called **megacities**. There are 28 megacities currently, with the majority in Asia.

Some graphs will include projections: that is, they will estimate data for periods in the future. For example, this graph goes up to 2020. The projections may be very well researched and as reliable as possible, but they are not the same as 'real' data that has actually been collected.

Reasons for rapid urban growth in poorer countries

1 High rates of rural–urban migration.

2 High fertility rates and population growth in cities.

3 Most economic developments (jobs) concentrated in cities.

EXAM ALERT!

Some students wrote about the **reasons** why people move to the cities from rural areas, here they **explain** rather than **describe**.

> Students have struggled with exam questions similar to this – **be prepared!**

Worked example

tier F&H

Describe the advantages that urban areas have over rural areas in poorer parts of the world.
(4 marks)

Urban areas have a bigger range of jobs than in rural areas, where most people farm to feed themselves. City jobs are also better paid than jobs in the rural areas, even if the jobs are in the informal sector. Many areas in cities have safer water, electricity supplies and sanitation, which are often lacking in rural areas. There are many more opportunities to improve your life in the city, while rural life must often seem like a dead end. There are more opportunities to get welfare and medical help and greater access to schools.

Now try this

Name two factors that could encourage counter-urbanisation in rich countries.   **(2 marks)**

Inner city issues

The needs of people living in inner city areas of the UK have often been ignored, leading to many problems.

A simple model of city layout

Key
- ■ CBD (Central Business District)
- ■ Inner city
- □ Suburbs

| As suburbs developed, people and jobs moved out of the inner city | People who couldn't afford to leave stayed in the inner city | Social problems increased | Governments tried to solve social problems with inner city investment |

Ethnic segregation

New immigrants can often only afford to live in the inner city where housing is cheap because it is old and in need of repair. Immigrants will tend to cluster in the same area of the inner city. There are different reasons for this:

- People like to live with people who are like them.
- New immigrants can get support from their community and families.
- New immigrants do not feel welcome outside their community.

Strategies for promoting integration in inner city areas

- Providing English classes for non-English speaking residents.
- Aiming for a greater mix of different ethnic groups in an area.
- Developing ethnic areas as tourist attractions, e.g. Curry Mile in Manchester.
- Encouraging ethnic minorities to set up their own businesses.

Government strategies for the inner city

1980

Urban Development Corporations poured money into 13 UK inner city areas. There were major improvements to housing and infrastructure.

1990

City Challenge brought local government, developers and the local community together to regenerate inner city areas.

2000

A focus on sustainable communities as well as new housing and facilities. Efforts were made to bring communities together and reduce carbon emissions.

2010

Worked example
tier F&H

Outline a strategy that aims to reduce ethnic segregation and encourage multicultural cities. **(4 marks)**

A strategy to improve integration in Hounslow is called Hounslow Language Service. The education authority has a special languages team who work with students from different backgrounds so they can become bilingual: they get really good at speaking English and also they continue to improve the language their parents speak. This encourages multiculturalism because it doesn't just say everyone should only be good at English. It also gives the same value to their other language.

Now try this

tier F&H

Describe how a government scheme to improve a named inner city area was put into action. **(6 marks)**

Housing issues and solutions

Not enough new homes are being built where they are needed.

More single people are wishing to live on their own.

Population growth means there are more people needing homes.

Reasons for the UK housing crisis

Marriage break-up: both former partners want a house in the same area.

More people are wanting to live in the south-east of the UK, where there are more jobs.

Build more houses – but where?

It would be easiest and cheapest to build thousands of new homes on **greenfield** sites outside the city, but it is more **sustainable** to redevelop **brownfield** sites inside the city.

What new housing should be built?

Semi-detached and detached houses on estates

Affordable / local authority housing to rent

Worked example   tier **H**

Explain why housing needs are different in different parts of an urban area.   **(4 marks)**

When people are poor but work in the CBD, they need affordable housing close to their jobs because they cannot afford to commute long distances. Better-off people want to live in the suburbs, where houses are larger and social problems not so intense. But housing shortages mean there aren't enough houses to meet demand.

Building accommodation over shops in the CBD to prevent the CBD becoming a 'dead heart' at night

Renovating and updating large Victorian houses around the edge of the CBD to provide quality apartments

Redeveloping urban areas to provide homes

Renovating 19th century inner city terraced housing

Building estates of detached and semi-detached housing on the urban fringe

Now try this

 You should make **two developed** points to explain your answer.

Explain the difference between greenfield sites and brownfield sites.   **(4 marks)**   tier **F n**

Inner city challenges

Two issues (traffic congestion and the decline of the CBD) need to be addressed to make the urban environment a pleasant one.

Congestion solutions

Too many cars on the roads leads to congestion and pollution. Ways to reduce congestion and pollution would be to:

- widen roads to allow more traffic to flow more easily
- build ring roads and bypasses to keep through traffic out of city centres
- introduce a congestion charge
- introduce park and ride schemes
- encourage car sharing schemes
- have better public transport, cycle lanes, cycle hire schemes, etc.

Revitalising the CBD

City authorities need to keep the CBD attractive and welcoming by:

- pedestrianising shopping streets and keeping traffic out of the centre
- redeveloping open-air shopping streets into covered shopping malls
- providing alternative functions such as entertainment so that the CBD is busy in both the daytime and the evening.

tier F n

Worked example

A street in Cardiff's CBD

Study the photo of Cardiff's CBD above. Explain how this CBD has been revitalised to make it more attractive for shoppers. **(4 marks)**

The street has been pedestrianised so shoppers can stroll around without any traffic danger, fumes or noise. Trees have been used to create a relaxing environment and people have been employed to keep the area clean and clear, which adds to its pleasant atmosphere.

EXAM ALERT!

Make sure you only make points that are relevant to **this** photo. Some students listed more points than they need and didn't explain them.

Students have struggled with exam questions similar to this – **be prepared!**

Now try this

Give **two** reasons why the Central Business District often adds to the housing problems of an urban area. **(2 marks)**

tier H

Squatter settlements

In developing countries, where urbanisation has been rapid, migrants from the countryside have built their own squatter settlements to live in and created their own jobs in the **informal sector** (low paid or unofficial jobs, such as street sellers, waste collectors etc.).

Tiny, cramped houses with no toilets

Houses made of scrap materials

Houses are built illegally – the people have no legal right to the land

No electricity – unless it can be tapped illegally from pylons

It is cheaper to live in squatter settlements than in other parts of the city

They are close to economic opportunities in the city

Houses can also be shops or workshops

No paved road access

No sewage system, rubbish collection or clean water supply

The Mathare squatter settlement in Nairobi, Kenya

Effects on people's lives

In bad conditions

☞ **Infectious diseases** – poor sanitation means disease spreads quickly and easily.

☞ **Floods** – squatter settlements are often built on land that floods frequently.

☞ **Pollution** – settlements may develop near to industry.

In better conditions

☝ **Employment** – increased chance of regular work.

☝ **Opportunities** – more than in countryside.

Worked example

tier F n

Explain how rapid urbanisation leads to the development of squatter settlements. **(4 marks)**

Rapid urbanisation is usually fed by rural–urban migration. People come to the cities but don't have any job or anywhere to live. They build somewhere to live and make a job in the informal sector. Also, the speed of urbanisation is too fast for city planners to keep up with. That is another reason why growth is haphazard rather than planned.

Now try this

What is meant by the term 'informal sector of the economy'? **(1 mark)**

tier F n

Squatter settlement redevelopment

Improvement strategies

- **Self-help** – local communities get together to improve things.
- **Site and service schemes** – authorities set up a site for squatter settlement dwellers and provide basic services.
- **Local authorities** – can be involved in different ways, including major redevelopments of whole area.

Case study

You need to know a case study showing how a squatter settlement has been redeveloped.

Case study

Sao Paulo

Location: São Paulo, Brazil's largest city. There are around 2500 favelas (squatter settlements), and many have grown up around new factories, along main roads and railway lines, including near the CBD and on floodplains.

Growth: for 40 years, poor rural migrants have been squatting in the favelas. Favelas were restricted to the outskirts, but they have spread to all unoccupied parts of the city. Two million people now live in favelas – 20% of São Paulo's population.

Problems: the favelas are often built on unsafe land and at first they have no electricity, piped water, sanitation, rubbish collection or roads. Gang crime is high and there can be infestations of cockroaches and rats.

Self-help: community groups have campaigned for improved services, better education and a reduction in crime. Small grants (micro loans) have been given to local groups to set up businesses.

Improvements: some of the older favelas (e.g. Paraisopolis) now have services. Some people are concerned that these improvements will encourage more people to the city.

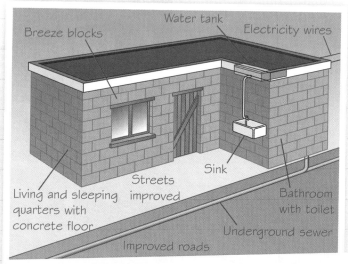

A self-help scheme house design used to improve squatter settlements in São Paulo, Brazil

Use the case study revision template on page 115 to organise the details you need for your revision.

Worked example   tier **Fn**

Explain the difference between self-help and site and service schemes.   **(2 marks)**

Self-help schemes come from the community helping itself, with some assistance from local authorities. Site and service schemes are when a new site with basic services is provided for squatter settlers by local authorities.

To do well, give the name and location of the squatter settlement and use detailed information to describe the redevelopment.

Now try this

With reference to a named example, describe how a squatter settlement has been redeveloped.
(6 marks + 3 marks SPag (F), 8 marks (H))

tier **F&H**

Rapid urbanisation and the environment

Rapid urbanisation and industrialisation cause environmental problems that are difficult to manage.

Big cities need a lot of water. Many cities are seriously depleting nearby water sources.

Smog caused by air pollution in Mexico City

The huge urban populations of a city create vast amounts of waste each day. Some of this is toxic and can't be disposed of easily.

Most big industry in poorer parts of the world is concentrated in cities. The industries cause most of the air and water pollution problems.

Air pollution in cities comes from burning fuel and vehicle exhausts as well as factory smokestacks. It can cause breathing problems for humans and acid rain.

Water pollution in cities comes from sewage, dumped waste from homes and chemical pollution by industries. It can harm human health.

Managing the problems

City authorities often do not have enough money to deal with problems. However, some cities are:

- developing new incinerators to deal with waste
- improving transport systems to reduce congestion
- implementing laws to reduce pollution from industry.

Worked example

tier **F n**

Name **three** sources of air pollution in a city undergoing rapid urbanisation.   **(3 marks)**

Low-grade petrol in cars; using wood as fuel for cooking; industrial air pollution from unregulated factories.

3-mark questions usually want you to do **three** things.

Now try this

Explain why it is difficult for city authorities in poorer parts of the world to dispose of a rapidly growing city's waste effectively.   **(4 marks)**

tier **F&H**

Sustainable cities

Sustainable urban living requires settlements that do not pollute the environment and which use resources in a way that means future generations can use them too.

Renovate old buildings to:
- enhance the appeal of the area
- provide visitor attactions.

Use brownfield sites for development to:
- improve appearance of these areas
- reduce loss of greenfield sites in the countryside.

Improve public transport systems by:
- linking bus, tram and rail routes
- providing feeder services to housing estates
- using environmentally friendly vehicles.

Ways to make urban living in the UK more sustainable

Building plans to include provision for **open spaces** to improve quality of life of urban residents.

Reduce waste by:
- recycling – 90% of household waste is recyclable
- reusing, e.g. bottles, plastic bags, etc.

Involve communities in local decision making:
- consult local people instead of imposing plans
- put people first; ask for and act on their ideas
- foster the growth of a community spirit.

Worked example tier F&H

Use a named case study to describe the features of a sustainable settlement.

(6 marks + 3 marks SPag (F), 6 marks (H))

Name of sustainable settlement: BedZED

BedZED is a sustainable settlement with around 100 houses and flats. It is carbon neutral, so it protects the environment. It was built on reclaimed land using natural and recycled building materials. A sustainable settlement should recycle waste, and BedZED residents recycle around 60% of their waste and also recycle wastewater. The site was built close to Hackbridge station so that residents would be able to use public transport, which is more sustainable than private cars. Residents also have a car sharing scheme for electric cars. Residents are included in decision making, and they don't have to live really green lifestyles if they don't want to.

BedZED an example of an eco-village in South London

 Make sure you use only relevant details from your case study rather than just writing down everything you can remember!

Now try this Make sure you can say which ones are connected to environmental issues and which are connected to a more sustainable society – social factors.

List the characteristics of your sustainable urban living case study that make it sustainable.

tier F&H

The rural–urban fringe

The rural–urban fringe is under pressure from urban sprawl.

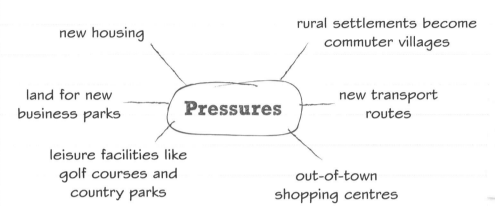

new housing

land for new business parks

Pressures

leisure facilities like golf courses and country parks

rural settlements become commuter villages

new transport routes

out-of-town shopping centres

The **rural–urban fringe** is an area around a city or town where urban land use mixes with rural land use and competes with it. **Urban sprawl** is the spread of urban areas into the surrounding countryside.

Reasons for development of urban fringe	Impacts of development
• Developers prefer rural areas on the edge of towns because there is more space to expand. Rural land is often cheaper. • People want to live in a pleasant rural environment and commute into work in the city. • Out-of-town shopping centres utilise cheaper land and locate near major roads to attract customers from all over the region.	• Traffic noise and development disturbs wildlife and changes the character of rural areas. • Traffic congestion increases as commuters travel into town. • The social and economic character of villages changes when they become commuter villages.

Commuter villages

Commuter villages are close enough to a town or city for people to travel to the city to work each day.

within commuting distance to a city

many young families

many older people who have retired from the city

Features of an expanding commuter village

richer newcomers and poorer locals

old core of village surrounded by new estates

wider range of shops and services

Worked example

tier **H**

Describe the changes that happen to a village as it expands and becomes a commuter settlement. **(4 marks)**

As a village becomes suburbanised, new houses are built within the village core and along routes into the village. Eventually new estates are built outside the core. Services in the village often change considerably. There are a lot more potential customers and a wider range of shops and services may develop. Commuter traffic puts pressure on rural roads. Conflicts can occur as a village expands: between old inhabitants and new, over house price increases, over increased traffic and parking.

Now try this   Try to make two points.

Describe how green belts attempt to prevent urban sprawl.   **(2 marks)**

tier **Fn**

Rural depopulation and decline

Rural areas are experiencing depopulation decline in services and a growth in second homes.

Rural depopulation

In very remote rural regions, depopulation is occurring as young people leave the area.

Case study

You need to know a case study of a remote rural area.

What is depopulation like in your case study area? Common case studies for this topic are Cornwall and Snowdonia.

low wages compared to urban jobs

rising poverty in rural areas

long hours and poor working conditions for many jobs

severe lack of jobs for most young people

Reasons

agricultural and food processing jobs are often seasonal

lack of affordable houses for young people

loss of services

isolation

Your case study should include information on a declining village. What has happened to the population structure, services and social life of your case study village?

Decline in rural services

- **Post Offices and shops.** It is expensive to keep small branches and shops open if few people use them.
- **Village schools.** Schools are expensive to run and it may be cheaper for local authorities to bus children to a bigger school in the nearest town.
- **Bus services.** It can be too expensive to run routes when most people own cars.

Second homes

Some people buy properties in remote rural areas as holiday homes. But they only live there for short periods each year. This:

- increases decline in village services
- inflates house prices, so local people can't afford to buy in the village.

Worked example   tier **F&H**

Use a **named example** to show the consequences of second home ownership in a rural area.   **(4 marks)**

Cornwall is a remote rural area in south-west England. In popular tourist areas like Port Isaac, 50–80% of houses are second homes. The demand for second homes among tourist visitors has pushed up prices so far that very few local people can afford them. Services have also changed, with many more very expensive restaurants for tourists and a lot fewer shops and services for local people.

A named example has to be **relevant** to the question.

Now try this   tier **F n**

Explain why the growth in second homes contributes to the decline in some rural services.   **(2 marks)**

Supporting rural areas

Rural areas are disappearing rapidly. Rural areas need to be protected and they need to be developed sustainably.

Conserving resources

The problem: Farmers need to make a living from their land. The more land they clear for farming, the more money they can make. But this would use up all the countryside's resources and would destroy wildlife habitats.

Solution: Environmental Stewardship Schemes. The government pays farmers to conserve the rural landscape.

Protecting the environment

The problem: Developers are another big threat to the rural environment. It is much cheaper and easier to build on greenfield sites.

Solutions:

- Green belt planning restrictions.
- Protection for special areas, for instance, National Parks, Areas of Outstanding Natural Beauty (AONB) etc.

Supporting the needs of the rural population

The problem: rural living is harder than urban living. Wages are lower, jobs and services are not as good, it can be isolating. It is hard to imagine rural life continuing like this – it is not sustainable.

Government initiatives: In the 1990s the government funded the **Rural Challenge**; 11 rural regeneration schemes shared £75 million to create jobs, train people for new skills, improve living standards and strengthen rural communities. 3000 jobs were created. There have been further projects since then; in 2009, a further £400000 was used to funded 90 community groups and businesses to help tackle poverty and increase inclusion in their rural communities.

Some sustainable solutions

- Help farmers diversify. This creates new types of jobs in rural areas. The Rural Development Programme funds this in England.
- Improve rural transport. Local councils sometimes fund schemes for this. For example, free buses to transport to college.
- New skills and training for rural communities. The Rural Challenge scheme does this.

Worked example

tier **F** n

Describe **two** examples of government initiatives to support the rural economy. **(2 marks)**

The Rural Development Programme gives funds to help farmers manage their land more sustainably. Rural Growth Network pilots in six areas are testing the best ways to promote sustainable economic growth in rural areas.

Now try this

What is farm diversification and how does it help support rural life? **(4 marks)**

tier **H**

Commercial farming 1

In some areas of the UK farms have become part of large businesses producing food for supermarkets and food processing factories.

Case study

You need to know a case study of a commercial farming area in the UK.

Agri-business

An agri-business is:

- is a large-scale farm. In some cases smaller farms have amalgamated to form one large commercial farm
- may be run by a large food company
- uses high-tech machinery, fertilisers and pesticides to achieve high quality produce that they can supply in large volumes and very reliably.

Environmental impacts

Most modern farming has a major impact on the environment.

- 👍 Farmers work hard to maintain and conserve the rural environment.
- 👎 They also use pesticides which kill many useful insects as well as pests.
- 👎 Fertilisers and slurry run off the land and pollute water courses: **eutrophication**.
- 👎 Most farm machinery runs off diesel, a major CO_2 emitter.

Market demands

Large **supermarket chains** control the market. They either own farms or arrange strict contracts with farmers, requiring them to produce crops of a particular size and / or weight.

Food processing **factories** are another big market. They buy in huge quantities of food from the cheapest, most reliable sources.

Case study

East Anglia: a farming area

Natural advantages

- flat, easily worked chalky boulder clay which isn't too sticky when wet
- high sunshine totals, less than 750 mm of rainfall.

Human advantages

- near London
- good transport links
- agricultural services, e.g. education, research, transport, etc.

Worked example

tier F&H

What are the advantages and disadvantages for agribusiness of using GM crops?   **(4 marks)**

Advantages: GM crops give increased yields which means agribusiness can make more money. Some GM crops also stay fresh longer which reduces agribusiness transportation costs. Disadvantages: protestors oppose GM crops and might damage fields, losing agribusinesses money. Customers in some countries also distrust GM foods, so agribusinesses would be concerned about their public image.

Use the case study revision template on page 115 to make sure you have all the detail you need.

Now try this

tier Fn

What is the advantage for agri-businesses of having very big farms?   **(2 marks)**

Commercial farming 2

There are different ways in which people are trying to reduce the impact that farming has on the environment.

Government policies

Government policies aim to reduce the environmental impacts of intensive commercial farming. Under the **Environmental Stewardship Scheme**, farmers get paid for doing specific things which help the environment, such as wildlife conservation, protecting natural resources, promoting public access and understanding of the countryside.

Nearly 70% of England's farmland is involved with ESSs or older schemes.

Case study

In your case study of commercial farming you also need to know about government policies to reduce impacts of farming and the development of organic farming.

Global economy

EU **tariffs** mean that food imports into the EU from poor countries are more expensive than products grown by EU farmers. Without these tariffs, many EU farmers would struggle against global competition and environmental stewardship would be affected.

Organic farming vs intensive farming

Intensive farming	Organic farming
High use of pesticides.	Pesticide use restricted. Wildlife encouraged to control pests naturally.
Artificial fertilisers.	No artificial fertilisers. Farmers use crop rotations, organic manure and clover crops to fix nitrogen in the soil.
Agri-businesses compete to have the highest quality and lowest priced products.	Organic food products are not perfect to look at and tend to be priced quite high.
Antibiotics used to keep livestock healthy and animals are confined to boost their weight gain.	Antibiotics are not used with livestock; instead traditional methods of disease control are used. Animals are free range.
High yields (artificial fertilisers and pesticides).	Lower yields (natural fertilisers and no pesticides).
Highly mechanised, so needs fewer people.	Low mechanisation, so needs more labour.
Large carbon footprint (mechanisation, air miles transporting the produce, etc.).	Lower carbon footprint – but organically reared livestock produce a lot more methane, a potent greenhouse gas.

Worked example

tier F n

Describe three methods farmers could use to reduce the impact of modern farming on the environment. **(3 marks)**

Replacing hedgerows provides habitats for wildlife; planting shelter belts helps reduce soil erosion by both wind and water; controls on release of animal slurry means that water pollution is reduced.

Focus your answer clearly on the extent to which organic farming may (or may not) **reduce** the environmental impacts of farming.

Now try this

tier H

Describe the advantages and disadvantages of organic farming as a way of reducing the environmental effects of farming. **(6 marks)**

Changing rural areas: tropical 1

Present day rural land use in tropical areas is less sustainable than traditional ways.

Non-sustainable methods

1 Cash crops industry

- Clearance of large areas of forest for soya growing or for cattle grazing.
- Many workers needed so traditional farmers are employed when they lose their land.
- Needs vast amount of water, which leads to disputes with traditional farmers.
- Mainly owned by multinational companies.

2 Forestry

Involves clearance of huge areas of forest for timber production.

3 Mining

Land cleared for open cast mines.

Cash crops, forestry and mining are not compatible with traditional subsistence farming methods.

Problems associated with deforestation – soil erosion

This soil erosion in Atlantic Forest, Brazil, is related to deforestation

Clearing the vegetation means the soil is unprotected from the heavy tropical train.

There are very few nutrients in tropical soils and the soil soon becomes exhausted.

Without vegetation, the tropical soils lose structure and are easily eroded.

Washed-away soil silts up rivers.

Fewer trees means there is less evaporation into the atmosphere which results in less rain.

Describe the impacts of mining in a tropical rural area.   **(4 marks)**

Most mining in tropical areas is open cast. Vegetation is cleared from the land. This removes protection from the soil and the heavy tropical rain causes soil erosion. Soil is washed away and goes into rivers, silting them up. Mining often uses lots of water. Pollutants are often washed from the mining area into the local environment.

 Make sure you concentrate on impacts and **not** causes or responses.

Now try this

Kenya has become the centre of flower farming for Western markets.
Explain the advantages and disadvantages of cash crops for Kenyan farmers.

(6 marks + 3 marks SPaG (F), 6 marks (H))

 tier F&H

Changing rural areas: tropical 2

The biggest changes to agriculture in tropical rural areas have come from irrigation and appropriate technology. Rural-urban migration also has an impact.

Irrigation

Irrigation of farmland in drier, sub-tropical areas has many advantages, but also some disadvantages.

Advantages

- 👍 Higher yields means more food is available to eat and to sell.
- 👍 Systems can be quite simple to set up and maintain.

> **Salinisation** is when high temperatures draw water and salts up through the soil by capillary action. The salt crusts on the soil surface and makes farming more difficult.

Disadvantages

- 👎 Poor water management can lead to **waterlogging** of the soil which can kill plants and **salinisation** which can ruin soil.
- 👎 Conflict can occur between countries when they are all using the same river for irrigation.

> **Waterlogging** A good drainage system is needed when land is irrigated to avoid the soil becoming waterlogged.

Appropriate technology

Appropriate technology means simple, low-cost technology that local people can use and repair themselves. Here are **three** examples that can improve farming.

 Water pumps powered by bicycle.

 Rainwater harvesting. This stores rainwater for use in the dry season.

 Universal nut sheller. This hand-powered machine really speeds up the process of shelling peanuts.

Rural–urban migration

The pressures on tropical rural areas, such as: soil erosion, salinisation, population growth and cash crops pushing farmers off the land, have forced people to leave for the cities.

It is usually young men who leave and farms are hard hit as there aren't enough people to work on them. At the same time, the booming urban populations need more and more food to eat but the farms struggle to meet demand.

Worked example

 tier H

Study this photo, which shows people in Eritrea constructing barriers to help prevent soil erosion. Explain why this is appropriate technology. **(4 marks)**

It is low-tech: something that everyone in the community can help build and fix. It is small scale: local farmers can put the walls where they need them. It fits with the local need to reduce soil erosion. It will also improve people's lives right now and also into the future.

> Make sure you use the evidence in the photo to help show your understanding.

Now try this

Explain how irrigation can lead to salinisation. **(2 marks)**

tier Fn

Measuring development 1

There are lots of different ways to measure development. Here are some examples:

Gross National Product (GNP)
Everything produced by a country in one year

Gross National Income (GNI) per head
Everything produced by a country in a year divided by the population of the country

Human Development Index (HDI)
A complex indicator made up of life expectancy, literacy, education and standards of living

Birth rate
The number of live births per thousand of the population per year

Death rate
The number of deaths per thousand of the population per year

Different measures of development

Infant mortality rate
The number of babies that die under 1 year old per thousand live babies born

People per doctor
The average number of people for each doctor in the country

Literacy rate
The percentage of adults who can read and write

Access to safe water
The percentage of people with access to safe drinking water

Life expectancy
The number of years remaining at a given age, e.g. at age 15, life expectancy is an additional 60 years (total age 75)

Limitations

GDP and GNI give an average picture across a country, so if there is a rich elite and a lot of very poor people this would not show up. At a national level, none of the indicators shows if there are differences between different parts of the country, e.g. rural and urban.

Correlations

The measures work best when one measure can be **correlated** with another, for example if countries with a low GNP also show limited access to safe water or high infant mortality. Correlations between measures are often analysed with **scatter graphs**.

Make sure you can interpret **scatter graphs**: see page 107 for more on scatter graphs.

Worked example

tier F n

Study this table which shows HDI ranks and figures for four countries.

Rank	Country	HDI 2011
1	Norway	0.943
50	Romania	0.781
100	Fiji	0.688
150	Cameroon	0.482

What does HDI measure? **(2 marks)**

HDI measures human development and ranks countries into order according to their HDI score. It combines statistics on life expectancy, literacy, education, and standards of living for each country.

Now try this

Study the table on this page showing HDI scores for different countries. Explain whether you would expect the GNI per head for each country to correlate with the HDI scores. **(2 marks)**

Measuring development 2

There are different ways of measuring global development.

The North–South Divide (1981)

The problem with this classification is that there are lots of differences within the North and South that this doesn't identify.

Four levels of income

The World Bank divides countries into high, upper-middle, lower-middle and low incomes. But this has problems: some low-income countries are rapidly industrialising, while some are still very poor.

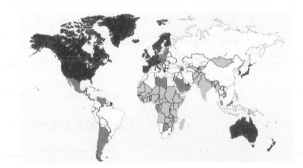

Human Development Index

More detailed classifications, such as the HDI map, are not as simple to use but do capture more of the differences within categories. They recognise social variations rather than being wholly economic.

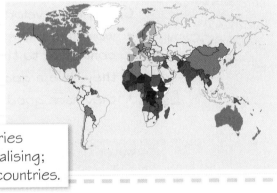

The **fivefold division**, based on wealth, divides countries into Rich industrialising; Oil-exporting; Newly industrialising; Former centrally planned and Heavily indebted poor countries.

Quality of life and standard of living

- Standard of living is an **economic** measure. Does a person have enough money to live on?

- Quality of life is a **social** measure. Do people have a long and healthy life?

  - Rich world perceptions of quality of life are different from poor world perceptions.

  - Rich world perceptions may overlook the attempts by people in the poorer parts of the world to improve their own quality of life.

Worked example

Explain why it is no longer considered valid to classify countries into first world, second world and third world countries. **(2 marks)**

First, it does not account for the wide variations in development within categories. Second, it assumed that the countries in the 'First World' were the best, superior to those in all other categories.

Now try this

'What is meant by...' means give a **definition**.

What is meant by the term 'quality of life'?   **(1 mark)**

What causes inequalities?

Development is affected by different factors. This can lead to global inequalities.

Environmental

Countries find it more difficult to develop when they:

- are landlocked and can't benefit from trade by sea
- are blighted by tropical diseases which affect people's ability to work
- have poor soils and arid conditions
- suffer from natural hazards, e.g. earthquakes, and do not have the money to repair the damage caused.

Economic

Countries find it more difficult to develop when:

- global trade favours developed countries
- tariffs make trade more expensive
- they produce mainly primary products which don't make much money
- they are in debt and spend money on interest payments rather than development.

Factors leading to global inequalities

- One in nine people in the world do not have access to safe water.
- Approximately 2 million people die from water-related illnesses every year.

- Access to safe water kick-starts development – people are no longer too ill to work, children can go to school, women can work instead of spending all day fetching water, and there is enough water to grow crops.

Social

Countries develop more easily when:

- women are able to work and contribute to the economy
- they have a good education system
- they have good water supplies.

Political

Countries find it difficult to develop when:

- political instability scares off investment
- wars disrupt the economy, destroy the infrastructure and cause displacement of people
- corrupt governments tap development funds for themselves.

Worked example

tier **H**

Explain how debt can be an obstacle to development. **(4 marks)**

When poor countries are in debt, they use up money paying the interest on the debt rather than spending it on development projects. Debt can also be an obstacle in everyday life. When farmers have a poor crop they are often forced into debt to survive. Then what money they have goes on paying interest instead of educating their children, for example, or improving their farms.

Now try this

tier **F&H**

Explain why countries which largely rely on exporting primary products are at a disadvantage in world trade. **(4 marks)**

The impact of a natural hazard

Natural hazards can have a major impact on development in poorer countries.

Use the case study template on page 115 to help you organise your case study of a natural hazard.

Use the case study template on page 115 to help you organise your case study of a natural hazard.

Case study

You need to know a case study of the impact of a natural hazard on global inequalities.

Case study

The 2010 Haiti earthquake

- date: 12 January 2010
- magnitude: 7.0
- location: near Port-au-Prince, capital city
- 220 000 people killed, 300 000 people badly injured
- 1 million people made homeless
- 4000 schools destroyed
- 80% of government buildings destroyed
- 19 million m³ of rubble filled the streets of Port-au-Prince, blocking roads.

Impact on development

World leaders promised $4.5 billion to help rebuild Haiti, but by 2012 only half that money had been delivered.

- Donor countries have insisted their companies do the reconstruction work instead of Haitian companies.
- Half a million Haitians are still homeless.
- 40% of rubble is not cleared.
- Experts think it will be another 10 years before Haiti will show signs of real recovery.

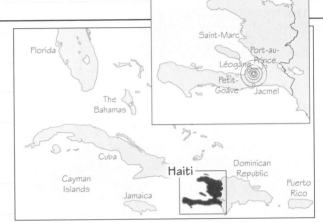

2012 compared to 2009

Haiti was one of the world's least developed countries before the earthquake and things haven't improved much since:

2009	2012
145th of 169 countries on the Human Development Index	158th on the HDI
GDP per head: $1300	GDP per head: $1300 (UK: $36 000)
Infant mortality rate: 59.59 per 1000	Infant mortality rate: 54.02 per 1000 (UK: 4.56)

Worked example   **tier F n**

Outline **two** physical factors that can reduce a poorer country's chance of development.   **(4 marks)**

Natural hazards like earthquakes make development in poorer countries difficult because they cannot always respond and rebuild as quickly as richer countries. Another physical factor is poor soils which can make it difficult for a country to grow food for itself, which can make development more difficult and increase global inequalities.

Now try this

Explain why it is often more difficult for a poorer country to recover from a natural disaster than it is for a richer country. Use a named example of a natural hazard in your answer.
(6 marks + 3 marks (F), 8 marks (H))

tier H

Is trade fair?

World trade has benefited developed countries rather than developing countries. There have been international attempts to address this.

World trade is out of balance

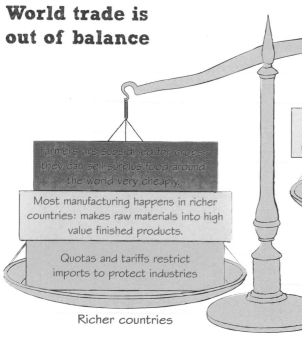

Farmers are subsidised for crops: they can sell surplus food around the world very cheaply.

Most manufacturing happens in richer countries: makes raw materials into high value finished products.

Quotas and tariffs restrict imports to protect industries

Richer countries

Farmers are not subsidised. They can't compete with cheap surplus products from rich countries

Many poorer countries depend on primary industries. They often have to import manufactured products from richer countries.

Poorer countries have to compete with each to offer the lowest prices for raw materials. Prices often fluctuate widely

Poorer countries

EXAM ALERT!

Your answer must be wholly based on Figure 1.

Students have struggled with exam questions similar to this – **be prepared!**

Balancing it out

International attempts to address the imbalance are:

- **Debt abolition** – richer countries can agree to cut the debt some poorer countries are burdened by.

- **Conservation swap** – the debt some poorer countries owe can be swapped for carrying out conservation projects.

- **Trading groups**, e.g. Fair trade – developing countries join a trading group where they can negotiate higher prices and set a fair price for products. Also, because they are working together they are not in competition so are less likely to undercut each other on price.

Fair trade agreements also aim to protect the **local environment** and **help communities** – making it sustainable development

Worked example

tier F n

Study Figure 1, which shows shares in world trade (exports) by region for 2009.

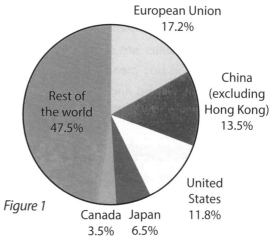

European Union 17.2%

China (excluding Hong Kong) 13.5%

Rest of the world 47.5%

United States 11.8%

Canada 3.5% Japan 6.5%

Figure 1

Explain **one** way in which **Figure 1** illustrates an imbalance in the pattern of world trade. **(2 marks)**

Half of all the exports in the world are produced by just four countries and a trading bloc. Apart from China, all these are rich countries. Trade is not equal.

Now try this

tier F n

Explain the difference between debt abolition and conservation swaps. **(2 marks)**

Aid and development

Aid can help close the development gap.

In disasters **emergency aid** is distributed on a short-term basis. **Development aid** is used for long-term projects.

Case study

You need to know a case study of a development project, its advantages and disadvantages.

Types of aid

- Emergency aid and development aid
- Government aid
  - Bilateral – aid direct from one country to another
  - Multilateral – aid to international agencies, e.g. World Bank
- Non-government aid (NGO)

Some donor countries place conditions on how the money is used, this is called **tied aid**. When there are no restrictions it is called **untied aid**.

Advantages and disadvantages of aid

For donor countries:

- 👍 Charity donations make people feel they are helping.
- 👍 Governmental aid can bring countries closer together and build alliances.
- 👎 People who give to charity can sometimes feel their aid is wasted.
- 👎 Feeling that charity should begin at home.
- 👎 Charity-giving fatigue.

For recipient countries:

- 👍 Emergency aid helps save lives.
- 👍 Development aid builds new skills, creates jobs and improves health and education. Often encourages sustainable approaches.
- 👎 Emergency aid sometimes doesn't provide what is actually needed.
- 👎 Tied aid makes the recipient country reliant on the donor country.
- 👎 Large-scale projects benefit big companies and urban areas rather than most local people.

Worked example

With reference to a named example, describe how a development project is encouraging sustainable development. **(8 marks + 3 marks SPaG)**

WaterAid is an NGO that helps people get access to safe water. One example of their work is Hamatabu village in Zambia. Women in the village had to spend up to 5 hours a day collecting water. With WaterAid providing technical help, the community dug a new well and a hand pump was installed. The community collects money each month to pay for any repairs. As a result of the project, illness from unsafe water has really reduced and women have time to farm their plots and to make craft items to sell. There is water for the crops and also for latrines. This development is sustainable because it involved the community and directly benefits the community. Water will help improve the village now and also well into the future.

Now try this

Make sure your answer is detailed and is based on your case study, **not** on generic information.

Explain the difference between bilateral aid and multilateral aid. **(2 marks)**

Development and the EU

The EU aims to reduce development differences between its 27 member countries.

Contrasting EU countries: rich and poor

Make sure you revise the two contrasting EU countries that you learned about.

Germany
- GDP per head: $39 100.
- Population: 81.1 million.
- 3.5 infant deaths per 1000 population.
- Life expectancy at birth: 80.2 years.
- Population below poverty line: 15.5%.

Bulgaria
- GDP per head: $14 200.
- Population: 7 million.
- 16.1 infant deaths per 1000 population.
- Life expectancy at birth: 73.8 years.
- Population below poverty line: 21.8%.

Germany is the EU's richest country. It is an industrial and exporting powerhouse, making the biggest contribution to EU budget and is at the centre of Europe.

Bulgaria is a recent member of the EU (joining in 1995). A former Communist bloc country it is one of the EU's poorest members on the periphery of Europe. It has problems with corruption in government.

How the EU reduces inequalities between countries

All EU countries pay 1% of their total wealth to a central budget and then a third of the total budget is spent on regional aid for less developed EU countries. The different funds are:

Structural Fund

The European Regional Development Fund

Improves investment and infrastructure.

This includes the **Urban II** fund: sustainable development projects for cities.

The European Social Fund

Pays for education, training and job creation.

The Cohesion Fund

Improves the environment and transport. Supports the development of renewable energy. Only for countries with living standards that are less than 90% of the EU average, inc. Bulgaria.

The Common Agricultural Policy (CAP)

Pays subsidies to all EU farmers; favours countries with big agriculture sectors.

The EU debt crisis

Major problems with the economies of some countries on the periphery of the EU put enormous strain on the ability of the EU to hold together. The richest countries, e.g. Germany, have had to bail out the poorer countries.

Worked example

tier F&H

With reference to a named example, give **two** reasons why a country in the EU has a living standard below average for the EU. **(4 marks)**

When Bulgaria was a Communist country it had very little investment into its economy, which is mainly rural. It is also on the periphery of Europe, a long way from the big market centres.

EXAM ALERT!

Some students wrote about Eastern Europe when the answer requires a **country**.

Students have struggled with exam questions similar to this – **be prepared!**

Now try this

tier F&H

Explain how the EU's Common Agricultural Policy helps farmers in poorer EU countries. **(2 marks)**

Going global

Globalisation is a process: how the countries of the world are getting more connected.

Globalisation – what is it?

Economic interests and a desire to make profits has encouraged companies from developed countries to produce in countries where labour is cheap and then to sell the products all over the world.

→

This pulls countries together in a global economy (globalisation).

→

Trade connections between countries has led to **interdependence** between them. Countries trade because one country has something the other country doesn't.

Technological developments and globalisation

Containerisation makes shipping goods cheaper and faster

→

Companies can make products cheaply in one country and ship them to sell in other countries

Satellites and submarine cables give superfast communication

→

Companies can communicate and make decisions globally

Impacts of globalisation

Overseas call centres

Call centres deal with customers on the telephone, by email or by messaging.

Companies in richer countries have moved operations to poorer countries because:

- wages are generally lower
- internet technology means cheap and clear international calls
- most people speak English in countries such as the Philippines and India where there are many call centres.

Localised industrial regions

Interconnected industries can cluster together in one area but still have global connections due to ICT and transport developments. An example is **Motorsport Valley** near Silverstone in Northamptonshire. There are nine Formula One companies based here.

Heathrow and Birmingham airports nearby for export and travel

M40 and MI nearby

Near to Silverstone race track for testing and events

Motorsport Valley location benefits

Beautiful countryside so pleasant place to live

Near to research facilities

Near to pool of highly-skilled workers

Worked example

What is 'globalisation'? **(2 marks)**

Globalisation is the way different countries are more and more connected to each other. These connections make them more interdependent.

Now try this

Describe the advantages to a company of moving its call centre operations from the UK to India.
(4 marks)

TNCs

Transnational corporations (TNCs) are major companies that operate in more than one country: for example, a company might have its headquarters in Switzerland, produce its goods in China, and sell in the US and Europe.

Case study

You need to know one case study of a TNC, including its advantages and disadvantages.

Advantages and disadvantages

TNCs aim to maximise their profits. Poorer countries may be able to benefit, but generally TNCs are not interested in helping them develop.

Advantages of TNCs in developing countries	Disadvantages of TNCs in developing countries
• TNCs bring money, modern technology and skills to poorer countries. • The country's infrastructure (roads, energy supplies, etc.) is improved by or for TNCs. • TNCs create jobs so people can buy more and pay more tax. • TNCs in manufacturing industries create exports for the country, so it can earn foreign currency. • The multiplier effect – other industries can grow up around TNC factories and supply them with parts.	• TNCs generally pay low wages and expect long hours. • Jobs are often boring, repetitive and don't develop many skills. • Most of the profits go back to the TNC's base in a rich country. • Generally they are not interested in improving working conditions any more than they have to. • Often they are not subject to health and safety and environmental regulations. • TNCs can leave a country if somewhere else becomes more profitable.

Case study

Tesco

Tesco is a TNC with UK headquarters.

- Tesco's strategy is to sell a wide range of products at cheap prices.
- It sources cheap products from around the world, including countries such as Bangladesh, Sri Lanka and Kenya.
- It sells its products in over 6000 stores in 13 countries, including India and China.
- There are 3146 stores in the UK and another 3638 worldwide.

Worked example

Describe how TNCs influence global interdependence. **(4 marks)**

TNCs are companies that operate in more than one country. TNCs generally have their headquarters in richer countries but locate their production in poorer countries. That makes the poorer countries dependent on the TNCs for jobs and investment. At the same time, TNCs rely on poorer countries for their profits, because they can make things cheaply and then sell them for a lot more money. Also TNCs increase global interdependence because they sell their products globally. All round the world people wear the same fashions, watch the same films, eat and drink the same products.

EXAM ALERT!

You must write about dependance **not** about the disadvantages of TNCs.

Students have struggled with exam questions similar to this – **be prepared!**

Now try this

Explain how the arrival of a TNC can have disadvantages for poor countries. **(4 marks)**

Manufacturing changes

Some parts of the world have seen rapid industrial growth while other parts of the world have experienced deindustrialisation.

How TNCs have affected global manufacturing

TNCs look for the cheapest place to make their products.

- They move production to factories in NICs – boosts **industrialisation**.
- They close down factories in HICs – increases **deindustrialisation**.
- More factories develop to supply the new production in the NICs.
- HIC factories cannot compete with TNC prices – they either relocate or shut down.

Key terms:

- **Deindustrialisation** – when manufacturing industry declines and becomes much less important
- **NICs** – newly industrialising countries
- **HICs** – highly industrialised countries.

Other factors affecting NIC industrialisation

Unions help workers in HICs to bargain or strike for better wages or conditions. Most NICs have no unions or only very weak ones.

HIC workers demand good pay and working conditions.

HIC governments tax companies to fund government spending.

HIC companies must obey strict laws to protect workers and the environment.

HIC Push factors

Emissions controls

Waste monitored

10% more pay

Official Strike

Government Health and Safety Executive

Movement of industry

NIC Pull factors

Opening hours: 0600-2000, Monday-Saturday

In NICs health and safety laws are often very weak, which means production can be faster and cheaper.

NIC workers will work long hours for low pay in order to get a job.

NICs set up tax breaks and other incentives to encourage foreign investment.

Worked example

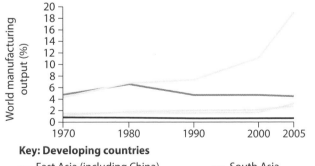

Study Figure 1, a graph showing shares of world manufacturing output between different parts of the developing world 1970–2005.

Which regions of the world have seen the biggest rise and smallest rise in manufacturing output? **(2 marks)**

East Asia (including China) has seen the biggest rise. Sub-Saharan Africa has not had any rise over this period at all, so it is the one with the smallest rise.

Key: Developing countries

— East Asia (including China)
— Latin America
— North Africa and Middle East
— South Asia (including India)
— Sub-Saharan Africa

Now try this

Explain the advantages for TNCs of locating manufacturing in newly industrialising countries (NICs). **(6 marks + 3 marks SPaG (F), 6 marks (H))**

China

Timeline

1949 People's Republic of China formed and first steel factory started production

1979 Foreign investment allowed in small areas

From the 1980s China's economy has grown rapidly and it is now an industrial giant

China's rise

1976 'Market socialism' reforms begin

1980 Special Economic Zones (SEZs) set up with tax incentives for foreign companies

cheap labour

working hour limits weakly enforced

Reasons for China's growth

health and safety weakly enforced

tax incentives and tax-free zones

restrictions on strikes and unions

Disadvantages of China's industrial growth

- Other countries find it hard to compete with China's low wages leading to deindustrialisation in HICs and NICs.

- Many Chinese are still poor because of low wages, 20% of the population live on less than $1 a day.

- The government has cracked down on moves by workers to gain democratic rights so working conditions are poor.

- China's rapidly rising demand for oil and other resources has helped push prices up around the world.

- There are high rates of air pollution in cities because there are few environmental regulations and 70% of China's rivers are polluted.

Worked example

Study this graph, which shows what the economic growth of China and the USA is expected to be in the future.

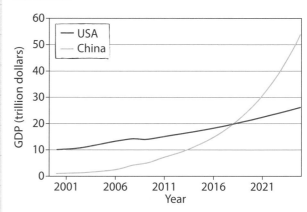

When is China expected to overtake the US as the world's largest economy? **(1 mark)**

The graph estimates China's GDP will overtake the US's GDP in 2018, when it reaches 20 trillion dollars.

EXAM ALERT!

Some students did not take enough care reading the graph and gave the wrong answer.

Students have struggled with exam questions similar to this – **be prepared!**

Now try this

In 1980, China spent $660 million importing oil. In 2011, it spent $227 billion importing oil.
Give **two** reasons that help explain this colossal increase in oil imports. **(4 marks)**

More energy!

The global demand for energy has gone up and up. This has had some serious impacts.

Rising demand for energy

World energy consumption increased 2.5% from 2010 to 2011.

The graph shows how almost all of the energy comes from oil, coal and gas: fossil fuels. Fossil fuels have a huge environmental effect because they release large amounts of CO_2 when burned.

This sort of graph is called an **area graph**. You find the value for each item in the graph by working out the difference between the top and bottom of its band for a particular year.

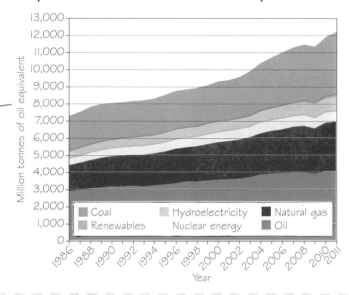

Million tonnes of oil equivalent vs Year

Coal | Hydroelectricity | Natural gas
Renewables | Nuclear energy | Oil

World population reached 7 billion in 2011. It is expected to peak at 9.5 billion by 2050. All these new people need energy too.

As countries get richer, their people buy more things which use more energy: air conditioning, cars, etc.

As new technology is developed, people want to buy new things or the latest version.

Impacts

Environmental
* More fossil fuel usage releases more CO_2; most scientists say this causes global warming.
* More risk of oil spills at sea, radioactive leaks from nuclear power stations, etc.

Economic
* Countries with energy resources get richer the more they sell.
* More energy means more economic development.

Social
* Climate change could mean millions of people migrating to escape drought.
* Gives more people the chance to access energy supplies, improving their lives.

Worked example

tier
F n

Describe one economic effect of an increasing demand for energy. **(2 marks)**

When demand pushes prices for fossil fuels such as oil and gas really high, it becomes economically worthwhile for companies to extract oil and gas from places that are difficult to work, for example deep parts of the sea or polar regions. Coal mines that had been shut down because they weren't making enough money can get reopened again.

Now try this

Explain why increasing wealth is linked to increasing demands for energy. **(4 marks)**

tier
F&H

Sustainable energy use

The use of renewable energy can help to achieve sustainable development. Renewable energy doesn't ever run out or damage the environment.

Case study

You need to know a case study of one type of renewable energy.

Types of renewable energy

Solar power – energy from the Sun.

Wind power – energy from wind.

HEP – hydroelectric power from running water.

Biomass – organic material processed to release gas.

Do not contribute to global warming, no carbon emissions

Inexhaustible, available for ever

Clean, no local air or water pollution

Widely available, one or more are likely to be available in most countries

Advantages of renewables

Locally available, many can meet small-scale needs, especially useful in developing countries

Can reduce globalisation costs, e.g. CO_2 emissions, pollution, fuel, etc.

Other ways to reduce the costs of globalisation

- Countries that beat their emissions targets get **carbon credits** which they can sell to other countries.
- Countries that help poor countries beat emission targets also get credits.
- Encouraging people to save energy by **recycling**.

International directives to cut carbon emissions

- **Kyoto Protocol** (1997): global agreement to cut emissions of CO_2 by 2012.
- **Copenhagen Accord** (2009): a new agreement that softened terms of Kyoto: countries asked to say what cuts they can make to CO_2 emissions by 2020.

EXAM ALERT!

Some students gave examples of renewable and non-renewable energy sources, but didn't explain the **differences**.

Students have struggled with exam questions similar to this – **be prepared!**

Worked example

tier F n

Explain the difference between renewable and non-renewable energy sources. **(2 marks)**

Renewable energy comes from sources that will not run out, such as the Sun, wind, biomass. Non-renewable energy comes from sources such as coal and oil: fossil fuels that will eventually all be used up.

Now try this

tier F&H

Explain why reducing, reusing and recycling would help reduce some of the costs of globalisation. **(4 marks)**

Food: we all want more

Around the world, there is an increasing demand for food. Attempts to meet this demand can have positive and negative repercussions.

Why is demand increasing?

Global population growth means more people need food. People in HICs want their favourite foods all year, not just when they are in season. In NICs, people are changing the sort of food they eat: more meat and exotic foods.

Also, extreme weather conditions (linked to global warming) cause damage to crops, e.g. droughts in US farming regions in 2012.

Positive and negative effects of the global demand for food

Social | Economic

✗ When farmers switch to intensive agriculture to produce food for export (cash crop) there is less food to feed local people.

✗ Farmers take on debts to pay for fertilisers, pesticides and insecticides in order to successfully farm intensively.

✗ If food prices go down farmers are ruined.

✓ If food prices go up farmers do well.

Environmental

✓ Intensive agriculture enables farmers to increase yields and grow seasonal crops all year.

✗ It uses a lot of energy to transport food to HIC countries so the carbon footprint of food production is high (food miles).

✗ In poorer countries people are forced to farm marginal land to meet local demand but this causes environmental degradation where soil fertility is significantly reduced by overproduction.

Political

✗ Disputes can occur over resources, e.g. water. For example, the course of Nile flows through many countries and farmers depend on the water for irrigation. If one country wants to take more of the water for their own use there is less water for countries further down stream.

Campaigns for locally produced food

The Campaign to Protect Rural England is one organisation that is trying to convince supermarkets to stock more local food and encourage people to eat foods only when they are in season in the UK, thereby reducing food miles and increasing revenue for local producers.

CPRE
Campaign to Protect Rural England
Standing up for your countryside

Worked example

tier F&H

Explain why the food miles of the food we eat in the UK now are often much higher than they were 50 years ago. **(4 marks)**

Fifty years ago, people in the UK ate food products that were mostly grown in the UK. However, people now want products that are not available in the UK or only available in certain seasons. For example, strawberries are available in summer in the UK, but people want them all year round. Supermarkets are able to source these foods and bring them to the UK at prices that customers are able to afford.

Now try this

What is environmental degradation? **(1 mark)**

tier F n

The tourism explosion

In the 1950s only about 20000 people travelled to another country for their holidays every year. Now it is close to one billion people each year.

Here are some of the reasons for the global explosion in tourism:

- More people in more countries can afford tourist trips.
- More places in more countries offer tourism.
- Technological advances such as affordable air travel and the internet make tourism cheaper, easier to organise and more enjoyable.

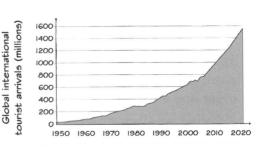

Different environments, different potentials

Different places attract tourists for different reasons. Here are three examples:

Coasts

Mountains

Cities

- sun, sand and sea
- activities include swimming snorkelling and sunbathing
- popular with young people, families and seniors

- natural beauty, peaceful locations and a physical challenge
- activities include walking, skiing and mountain biking
- popular with walkers, climbers and sightseers

- culture, entertainment and lifestyle
- activities include museums, galleries, shows and nightlife
- popular with young people and retired people

Economic importance

Tourism is one of the world's biggest industries. It is worth 3 billion dollars a day.

Tourism creates jobs and benefits businesses like hotels and restaurants.

Some countries like Thailand and Jamaica are very dependent on income from tourism.

Tourism doesn't benefit all regions of a country – half of all tourist money in the UK is spent in London.

Worked example

Explain how tourism can help a country to develop economically. **(2 marks)**

Tourism can really benefit a country's economy because it is labour intensive, which means it creates lots of jobs. Tourism creates lots of jobs in hotels and tourist shops and attractions.

To give a developed answer you should make two clear points.

Now try this

Give **two** reasons why tourists might want to travel to a city destination. **(2 marks)**

Tourism in the UK

The UK is the world's seventh largest tourist destination. Here are four external factors which affect how many tourists come to the UK each year.

1 Economy

👍 If a country's economy is doing well, its people have more money for trips to the UK

👎 If a country is not doing so well, its visitor numbers to the UK will drop

2 Exchange rates

👍 A low £ exchange rate makes the UK cheaper for visitors from other countries

👎 A high £ rate makes the UK expensive to visit

External factors

3 Security

👍 No recent terrorist incidents make the UK seem a safe place to visit

👎 'Security issues' make the UK seem a potentially dangerous place to visit

4 Media coverage

👍 Positive media coverage of UK makes lots of people want to visit

👎 Negative media coverage of UK stops lots of people wanting to visit

---

Worked example

 tier **Fn**

Study Figure A, which shows top visitor attractions in the UK in 2011.

Figure A

Top attractions in 2011	Number of visitors
British Museum	5.8 million
National Gallery	5.3 million
Tate Modern	4.9 million
Natural History Museum	4.9 million
Science Museum	2.9 million

Use Figure A to suggest why many tourists come on holiday to the UK.   **(2 marks)**

All the top attractions in Figure A are museums or art galleries so lots of tourists must come to see culture and history. The UK offers tourists lots of cultural attractions to visit.

UK tourism statistics

✓ 30.7 million tourists visited the UK in 2011 and this is set to grow.

✓ In 2009, tourism was worth £115 billion to the UK – nearly 9% of the whole economy.

✓ About 2.6 million people work in tourism in the UK.

✓ Most people who come to Britain are from France, Germany and the USA.

✓ UK residents spend £13 billion a year on trips within the UK.

EXAM ALERT!

Make sure you refer your answer to Figure A. **Describe** what the table shows, then **explain** what that tells you about why tourists might visit the UK.

Students have struggled with exam questions similar to this – **be prepared!**

Now try this

The photo shows a major sporting event in the UK in 2012. Give **two** reasons why this event could increase visitor numbers to the UK.   **(2 marks)**

 tier **Fn**

UK tourism: coastal resort

Blackpool is a UK coastal resort in north-west England.

Blackpool is a good example of a UK coastal resort, but revise details for the case study you have done at school.

Case study

You need to know a case study of either a UK coastal resort or a National Park. Page 99 is about a National Park, and this page is about a coastal resort – you don't need to revise both!

Blackpool: growth and challenges

Tourist numbers kept growing in Blackpool till the 1960s

Greetings from Blackpool

Greetings from Spain

In the 1970s, package tours to Mediterranean resorts hit Blackpool hard. Visitor numbers halved from 20 million in the 1980s to 10 million by 2007. Investment fell and the resort began to look shabby.

The resort life cycle model explains what often happens to tourist numbers over time

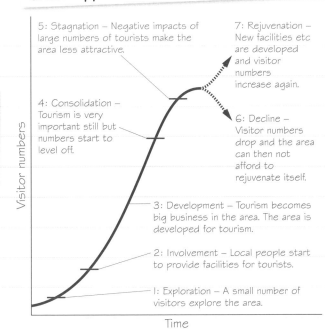

5: Stagnation – Negative impacts of large numbers of tourists make the area less attractive.

7: Rejuvenation – New facilities etc are developed and visitor numbers increase again.

4: Consolidation – Tourism is very important still but numbers start to level off.

6: Decline – Visitor numbers drop and the area can then not afford to rejuvenate itself.

3: Development – Tourism becomes big business in the area. The area is developed for tourism.

2: Involvement – Local people start to provide facilities for tourists.

1: Exploration – A small number of visitors explore the area.

Visitor numbers

Time

Blackpool now

In 2010 and 2011, visitor numbers to Blackpool increased to around 12 million a year. This increase is due to the following strategies to revive the resort.

- Facilities have been improved – old buildings have been pulled down, beaches cleaned up and the Blackpool Illuminations transformed with £10 million of new investment.

- New attractions have been added, such as Water World, however, a plan to open a Las Vegas-style super-casino failed.

- New types of visitors have been encouraged to visit – Blackpool has marketed itself as a venue for conferences and festivals, which can happen in the quiet winter months.

Worked example

tier F&H

Explain how external factors can affect the popularity of a tourist resort. **(4 marks)**

In the 1970s, tourism companies began offering package holidays to Mediterranean resorts. This external factor had a big negative impact on visitor numbers to UK coastal resorts like Blackpool. In 2008, the financial crisis meant fewer people went abroad for their holidays. This external factor increased the popularity of UK coastal resorts again.

Now try this

Give one characteristic of the 'rejuvenation' stage in the tourist area life cycle model. **(1 mark)**

tier Fn

UK tourism: National Park

The Lake District is a National Park in north-west England.

Case study

You need to know a case study of either a UK coastal resort or a National Park. Page 98 is about a UK coastal resort and this page is about a National Park – don't revise both!

Timeline

The Lake District – tourism 1800–2012

Early 1800s Poets write about Lake District and draw people's attention to its beauty

1851 Railways reach the Lake District

1951 Lake District becomes a National Park

1960s–1990s M6 built improving access to area

2012 16 million visitors a year

Lake District challenges

Lots of people visit the Lake District creating challenges. The table shows how strategies have been used to solve these problems.

You can use the case study template on page 116 to help you organise your case study revision.

Honey pot managmun

Challenges	Strategies
Traffic congestion on roads and parking problems	Park and ride schemes and promotion of car-free trips around the Park
Footpath erosion	'Fix the Fells' scheme organises footpath repair and fundraises for the money to do this
Second homes and holiday homes push up price of houses for local people	Development schemes encourage developers to build affordable housing for local people
Pollution and litter affecting the environment	Adverts, events, signs, talks and guided walks educate and persuade visitors to act responsibly

Plans for the future

- Encourage visitors to stay longer so they spend more on local businesses.
- Encourage visitors to use public transport rather than their cars.
- Expand the range of attractions to attract younger visitors.
- Keep natural areas and towns beautiful and welcoming.
- Involve more local people in tourism.

Worked example

What is the meaning of the term **seasonal unemployment**?   **(1 mark)**

Seasonal unemployment means that most of the jobs in an area are only available for part of the year.

Now try this

Outline how tourists can damage the environment of a popular tourist area.   **(2 marks)**

Mass tourism: good or bad?

Mass tourism is tourism on a large scale to one country or to one region.

Over 200000 Jamaicans work in tourism, so Jamaica is very dependant on the tourist industry.

Case study

You need to know a case study of a tropical area which attracts a large number of visitors. Make sure you know the positive and negative impacts of mass tourism in your case study area.

Economic impacts of mass tourism

Positive	Negative
👍 Brings money into the country's economy.	👎 Money often goes to foreign-owned holiday companies – this is called the leakage effect.
👍 Creates jobs for local people.	👎 Jobs for locals are often low paid and often seasonal.
👍 Brings new infrastructure to the region.	👎 Mass tourism can cause traffic congestion which affects local businesses.

Environmental impacts of mass tourism

Positive	Negative
👍 Can increase awareness of need for nature conservation areas.	👎 The local environment can be destroyed to provide tourist facilities.
👍 Tourism money can help pay to protect and repair the environment.	👎 Tourism creates local visual, water and air pollution problems.
👍 May actually help preserve areas or species, e.g., if there are trips to see sea life it is important the sea life is preserved so the tourists have something to see.	👎 Tourist activity can directly damage the environment. Air travel means massive CO_2 emissions.

EXAM ALERT!

Some students answered this question by saying what activities people did on holiday. The question is about the things that make mass tourism **different** from other types of tourism.

Students have struggled with exam questions similar to this – **be prepared!**

Worked example

Describe **two** features of mass tourism. **(2 marks)**

Mass tourism is organised tourism for large numbers of people going to the same place for their holiday.

Now try this

For a tropical tourist area you have studied, describe **two** positive impacts from mass tourism and **two** negative impacts. **(4 marks)**

Keeping tourism successful

For tourism to stay successful, negative impacts have to be reduced otherwise local people won't support tourism and tourist numbers will start to decline.

Strategies to reduce negative impacts of mass tourism

Help local people to benefit more from tourism

Tourists encouraged to be more responsible

Tourism companies encouraged to develop less damaging types of activities

Case study

Kenya's National Tourism Policy (2006)

Aim: double tourist numbers by 2020.

Strategies:

1. More money for Tourist Police to improve security for tourists.

2. Cultural tourism developed so tourists travel all over, not just beach and wildlife areas.

3. Game park ticket money used to protect the environment and shared with local people.

4. Holiday activities diversified: e.g. river activities, climbing and hiking.

5. New roads and airports built across Kenya.

Worked example tier F&H

Kenya tourism Fact File

- tourism = 15% of Kenya's economic production
- 250 000 people employed in tourism
- 2.4 million people visited Kenya's 27 parks and game reserves in 2009

Using **Figure 1** and your own knowledge, explain why a reduction in tourism would have negative effects on Kenya's economy. **(4 marks)**

Kenya's dependence on tourism is high as it represents 15% of the country's total production. 250 000 people are directly employed in tourism so if people lost their jobs they would not pay taxes to the government. Also, these people would have less money to spend which would have a negative impact on Kenya's economy.

Now try this tier H

Outline the strategies in your tropical tourist case study area for maintaining the importance of tourism and reducing its negative effects.
(6 marks + 3 marks SPaG (F), 8 marks (H))

 Make sure you outline **both** the strategies for maintaining the importance of tourism and for reducing its negative impacts.

Extreme tourism

Travelling to extreme environments, such as Antarctica, is becoming more popular. But extreme environments are very fragile, so the risk of tourists damaging them is high.

Case study

You need to know a case study of an extreme are and the extent to which it can cope with a tourist industry.

keen to have a really adventurous holiday

something totally different, new experience

rejection of mass tourism destinations

air travel makes destinations easier to get to

wildlife that can't be seen anywhere else

Attractions of extreme environments

keen to visit environments that may disappear or change

technological improvements (e.g. clothing) so tourists can visit in comfort

more tour operators offering extreme environment destinations

more people globally can afford the expense (retired people)

Case study

Antarctica

Why extreme?

Antarctica is at the South Pole and 98% of it is covered in ice. It is the coldest place on Earth with temperatures regularly down to minus 30°C.

Why popular?

Since the 1990s, the number of tourists visiting Antarctica has grown from around 7000 to 45 000.

Reasons for this growth:

- extremely remote and exciting
- unspoiled wilderness: no humans
- wildlife: penguins, seals, whales, etc.

Impacts of tourism

- Tramples fragile plant life.
- Litter will not degrade for centuries.
- Cruise ships and tourists disturb wildlife.
- Cruise ships may leak fuel, oil or sewage which can damage marine life.

Worked example

Describe **two** ways in which a **named** extreme environment is being protected from the impact of tourists. **(4 marks)**

Antarctica is an extreme environment that is becoming more and more popular with adventure tourists. The Association of Antarctica Tour Operators limits the numbers of people allowed off cruise ships, which reduces trampling damage and disruption to wildlife. It also forbids cruise ships dumping any wastewater. This reduces the risk of pollution. But these are guidelines and not all tour operators have signed up to them.

Now try this · tier F&H

Use examples that are specific to your named extreme environment.

Describe the types of activities that take place in the extreme environment you have studied.

(6 marks + 3 marks SPaG (F), 6 marks (H))

Ecotourism

Ecotourism is necessary to help make some tourist areas sustainable.

minimise the impact of tourism on the local environment and culture

educate tourists about how to act responsibly

inform local people about why their environment is important

Ecotourism should...

ensure tourists do not damage the local environment

benefit local people by creating jobs and trade opportunities

use resources sustainably

help to conserve the local environment

Case study

The Amazon and ecotourism

Benefits

- Tourists live in ecolodges which are built from local materials. The lodges serve local food and are run by local cooperatives. They use renewable energy and recycle water.

- Local people sell local products and some act as guides. Tourists can also learn crafts and skills from local people.

Why conservation and stewardship are needed

- Rainforest soils are fragile and without vegetation the soil washes away and flooding increases downstream.

- Rainforests soak up CO_2 so protecting them helps reduce the impact of rising greenhouse emissions.

- Rare animals and plants need to be protected to avoid extinction.

- The biodiversity of the rainforest needs to be preserved, lost species could have provided important medical cures

- Without the rainforest, indigenous people cannot follow traditional ways of life.

Worked example

Explain why sustainable development is an important goal of many ecotourism projects.
(4 marks)

When ecotourism minimises damage to the local ecosystem it is ensuring that this ecosystem can be used sustainably. When local people are involved in running an ecotourism project, this is sustainable because they will want their children to be able to benefit from it, too. When an ecotourism project is built using local skills, this is sustainable because local people can then make repairs.

Name an example of ecotourism. Include the country it is located in.
(1 mark)

Stimulus materials – an introduction

Some of the questions in the exam will have **stimulus materials** to use. These could be, for example, maps or graphs, tables, a text extract, a photo or a cartoon. They are never there just to look pretty – you need to use them correctly if you want to do well.

Maps are regularly used as stimulus materials for questions. You need to use your map skills when you tackle them.

The key thing to remember when using stimulus materials is **not** to just repeat the information that the stimulus material has given.

Instead you need to show that you have used the material so that it answers the question.

This is a **topological map** with very little detail on it showing the important things for you to answer the question. Even simple maps like this will have a scale, a key and a compass indicating north. Make sure you use all three as appropriate in your answer.

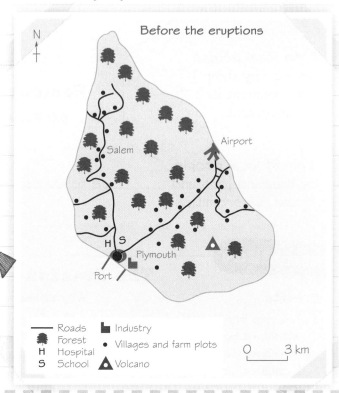

Before the eruptions

Key:
— Roads
🌲 Forest
H Hospital
S School
■ Industry
• Villages and farm plots
▲ Volcano

0 ——— 3 km

Worked example   tier H

An American company has operated the Mount Ertsberg gold, silver and copper mine in Indonesia for over 20 years and has turned the mountain into a 600-metre hole. The mine generates 700 000 tons of waste a day and this now covers about 3 square miles. This waste produces levels of copper so high that almost all fish have disappeared from nearly 90 square miles of wetlands downstream from the mine.

Using this extract from a newspaper article and your own knowledge, describe the impact of mining on an area of tropical rainforest.   **(4 marks)**

The creation of a 600-metre hole on the mountainside means that the forest has had to be destroyed and has caused visual pollution. The copper waste covers about 3 square miles and has killed all the fish downstream which has destroyed the local ecosystem because the river water is now poisonous.

Now try this

Examine the student's answer to the example exam question on this page. It has been colour coded for this activity. Which colour has been used for sections that the student has lifted straight out of the extract and which colour has been used for sections where the student has made a developed statement using their own knowledge?

Using and interpreting photos

You should be able to respond to and interpret ground, aerial and satellite photographs.

Different kinds of photographs

Ground-level photograph: shows lots of foreground detail. Use foreground and background to describe where things are in these types of photo.

Oblique aerial photograph: shows more of the area than a ground-level photo, and features are easier to identify than a vertical photo. But it is hard to judge scale for background features.

Satellite image: measures differences in energy reflected by different surfaces. False colour images convert this data into colours we can recognise. True colour images show us what the satellite sees, e.g. vegetation shows up red.

Vertical aerial photograph: these have a plan view, like maps. But details can be hard to identify.

The Five Ws

When working with photos, be sure to remember the Five Ws.

What does the photo show?

Why was it taken?

Who are the people in it?

Where was it taken?

When was it taken (to indicate how long ago it was taken, what time of day, etc.)?

Worked example

tier F n

Give **two** pieces of evidence from this photo from Rwanda that show how local people have overcome the physical difficulties of this area.

(2 marks)

Terraces mean farmers can grow crops on steep slopes. Leaving vegetation at the edge of each terrace helps prevent soil erosion from heavy rains.

Now try this

Describe **two** ways in which satellite photos like the one shown above can be used to monitor tropical storms.

(4 marks)

tier F n

Labelling and annotating

- Photographs and sketches are labelled and annotated in the same way.
- Only include the features that are relevant to the question.
- If you are asked to sketch something, draw clearly but don't worry about creating a work of art! Include a frame so that you can sketch within it.

This is a sketch of the photo above with labels and annotations

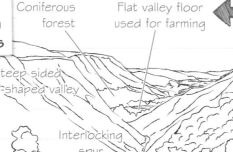

Coniferous forest

Flat valley floor used for farming

Annotations are sentences which offer **explanation**. Questions may require you to annotate photographs, sketches, maps, graphs or diagrams.

Steep-sided V-shaped valley

Labels are either one word or a short sentence which indicates what something is.

Interlocking spur

Deciduous woods

Meandering river

Worked example

Look at the OS map extract of Warkworth and Amble. Mark the following onto the sketch map:

1 A tourist information feature **(1 mark)**
2 A water feature **(1 mark)**

EXAM ALERT!

Some students lost marks because they did not locate the features carefully with the arrowhead actually touching the feature.

> Students have struggled with exam questions similar to this – **be prepared!**

Now try this

tier **F&H**

This photograph was taken at grid reference 272049. Mark with an X on the sketch map opposite the location of the lighthouse shown in the photograph. **(1 mark)**

tier **F n**

Graph and diagram skills

In your exam you may be asked to complete different types of graphs and diagrams. You may also be asked to interpret them which means describing and explaining what they show.

Line graph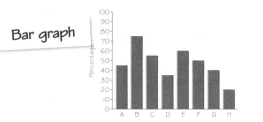

Line graphs are used to plot continuous data. They are often used to show how something varies over time. Make sure you plot the points accurately and join the points with a continuous line.

Bar graph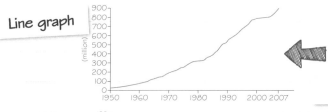

Bar graphs are used to plot discontinuous data. Make sure you draw bar graphs with a ruler to keep the lines straight.

Scatter graph

Scatter graphs show the relationship between two sets of figures. It is the pattern the points make that is important, so don't join up the points. If the line of best fit slopes downwards it is a negative relationship; if it is upwards it is a positive correlation. Some scatter graphs do not show any relationship.

Divided graphs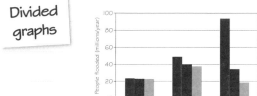

Divided graphs are useful for comparing data from one area or time with data from another. Divided graphs can also be drawn as one long bar divided up into segments like a pie chart.

Pie chart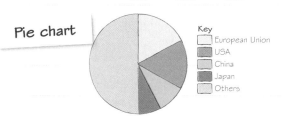

Pie charts show proportions. They are easy to read and fairly simple to put together. Data needs to be converted into percentages first and then into proportions of 360° – the whole pie.

Worked example

 tier **F**n

 Some questions will ask you to complete graphs and some to describe or interpret them.

Study the scatter graph shown above. Describe the relationship, if any, that it shows between access to improved water supply and life expectancy. **(2 marks)**

The line of best fit suggests a positive relationship, so as access to improved water supply increases, so does life expectancy.

Now try this

tier **H**

What kind of graph would you use to illustrate the following sets of data?

(**a**) Population growth in China from 1950 to 2010. **(1 mark)**

(**b**) The relationship between the size of settlements and the number of services in each. **(1 mark)**

(**c**) The proportion of people from different ethnic groups living in an inner city area. **(1 mark)**

Map types

You need to be able to complete different types of maps and may be asked to compare and interpret them too which means describing and explaining what they show.

Choropleth map

Key
■ over 75%
■ 50–75%
■ 25–50%
☐ 10–25%
☐ under 10%

0   300km

Choropleth maps are shading maps. They show how the value of something changes across the map. If you are completing a choropleth map, study the key carefully so you know what type of shading you need to use. Shading gets darker as values increase.

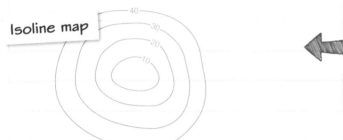

Isoline map

Isoline maps have lines joining points of equal value. When completing an isoline map, pay close attention to the size of the interval between the isolines already on the map and keep to it for any new lines you add.

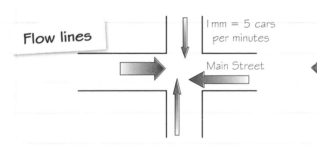

Flow lines

1 mm = 5 cars per minutes

Main Street

Flow lines show movements or flows. Some flow lines are proportional: the thickness of the line represents different ranges of data.

Study the key to see what thickness of line is used for what range of data.

Desire lines are a type of flow line map. They show journeys between two locations. Use the map scale to measure distances precisely.

Proportional map

Importers
Exporters
The figures show the percentage of the world total

Proportional maps use symbols drawn onto the map to show the proportions of something in different areas. The symbols can be anything, but are usually simple shapes such as circles. The larger the value, the larger the symbol.

Some questions will ask you to complete maps and some to describe or interpret them.

Worked example
tier **Fn**

What is a topological map?   **(1 mark)**

It is a simplified map which just shows the most important features or the most relevant features for a particular purpose.

Now try this
tier **F&H**

Study the proportional map shown above. What countries are the largest exporter and the largest importer of wheat?   **(2 marks)**

Describing maps

You need to be able to recognise and describe the distribution and patterns shown by the types of map found in **atlases**.

Satellite images and maps.

Political maps which show the outline of countries.

Maps which show the **distribution** of vegetation type, e.g. location of tropical forests.

Polar   Tropical
Temperature   Mediterranean

Climate zones which reflect global variations in precipitation and rainfall.

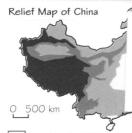

Relief Map of China

Relief maps showing the **height** and **shape** of the land.

0   500 km

Level 1: 0–1,150 feet
Level 2: 1,150–4,500 feet
Level 3: 4,500–8,200 feet
Level 4: 8,200–18,000 feet

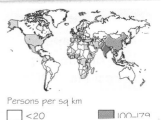

Persons per sq km
☐ <20   ☐ 100–179

Population distribution maps which show how **spread out** people are within an area.

Describing patterns

In the exams, you will be asked to describe a **distribution** or **pattern**. Use the letters GSE to help you structure your descriptions:

G – General overall trend or pattern

S – Specific examples that illustrate the trend or pattern

E – Exception – note any anomalies that do not fit in with the general pattern or trend

N

Key
■ Over 1500mm
■ 1000–1500mm
□ 625–999mm
□ Under 625mm

Average annual UK precipitation

0   160km

Worked example

tier F&H

Describe the distribution of rainfall shown by the map above.   **(4 marks)**

The general pattern shown on the map is a divide between high annual precipitation (over 1000 mm) in the western half of the British Isles and lower annual precipitation (under 1000 mm) in the eastern half. There are particularly high rainfall levels (over 1500 mm) in the regions of northern Scotland, north-west England, Wales and south-west England, and particularly low levels of annual precipitation (under 625 mm) in a belt running through Lincolnshire and Cambridgeshire down to the London area, although this belt also has areas of slightly higher annual precipitation near the coasts of Norfolk and Yorkshire.

Do not fall into the trap of explaining why the pattern happens when you have been asked to **describe**.

Now try this

Which two of the atlas maps at the top of this page would be most useful in explaining the global distribution of large-scale ecosystems?   **(2 marks)**

tier F&H

Comparing maps

You need to be able to compare two maps and compare maps with photos. You also need to be able to use maps at different scales.

> Be careful with your units when using OS maps at different scales, especially when you are measuring distances.

What is scale?

A map's scale tells you how much smaller the area shown on the map is compared to the area in real life.

- For OS maps at 1:25 000 scale, 1 cm on the map represents 25 000 cm in real life (250 m).

> Large-scale maps show a bigger area but are not very detailed.

- For OS maps at 1:50 000 scale, 1 cm on the map represents 50 000 cm in real life 500 m.

> Smaller-scale maps show the surface of the land in more detail, but can only cover a small area.

Worked example

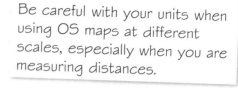

Study Figure 1, a 1:25 000 Ordnance Survey map extract of Bolton Abbey.

How far is it from the car park at Bolton Bridge along the B6160 to the car park at Bolton Abbey?

(1 mark)

1 kilometre

A question may ask for a straight-line distance or for a distance along a road, railway, etc.

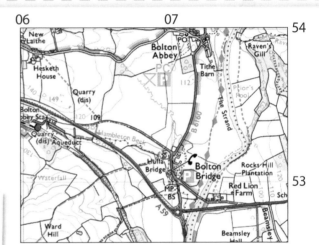

Figure 1

Comparing two maps

- Remember to talk about **similarities** and **differences** when comparing two maps.
- Look for patterns that might be significant: where things are and also where they are not.
- Always look to use the **scale** if possible, providing measurements of the things being compared.
- Use **grid references** to compare particular features.

Comparing maps and photos

There are some key differences.

- Maps (usually) have a scale and show permanent features in a landscape.
- Photos do not have a scale and show things only as they were at that moment.
- Aerial (vertical) photos and satellite photos are easier to compare to maps because they have a plan view; oblique photos and ground view photos are harder.
- Try to match up really obvious features on the photo to the map then you can orientate yourself.

Now try this

The map extract above is 1:25 000 scale. What would the distance between the two car parks be if it were a 1:50 000 scale?   **(1 mark)**

Exam skills

The most important thing to remember about exams is to read the question carefully. Make sure you have broken the question down so that you answer all parts of the question.

Command terms

Command terms tell you what to do with a question. Here are the most common ones.

Describe = write down what you can see.

Explain = give reasons for (why? / how?).

The most important thing you can do is to read the question properly. Underline the command words so you know what it is asking you to do.

List = make a list (could be a bullet list).

Compare = write down similarities and differences.

Contrast = write down the differences.

Outline = give a few details or write down the most important points.

To what extent ...? = come to a conclusion after giving different points of view.

Annotate = add labels with details.

Managing your time

Five top tips for managing your time in your exams.

1 Work out how much time you have for each question: use the number of marks to guide you.

2 Don't spend too long on the first question, which is usually only 1 or 2 marks.

3 Write concisely and to the point.

4 Don't waste time writing out the question: get straight on with your answer.

5 If you are running out of time on your last question, answer in note form rather than not writing anything.

But remember that this would affect your marks for spelling, punctuation and grammar.

Question types

Short answer questions

For questions worth 3 marks or fewer, try to make the same number of points as there are marks available.

Extended answer questions

For questions worth 4 marks or more concentrate on giving a really clear answer that sticks closely to what the question asks. Try to develop the points you make so they provide reasons, give details or identify points of similarity or difference, depending on what the question is asking for. Link your points together to build up your answer step by step.

A **basic answer**: answers will be simple and brief.

A **good answer**: statements are **clear** and linked together.

A **really good answer**: statements will be **detailed**, **clear**, **linked** and **well organised**. Remember also to consider your spelling, punctuation and grammar.

Now try this

'SPaG stand for speling punctuation and grammer.'
Can you find the five mistakes in the sentence above?

Exam skills: using case studies

Make sure you know how many case studies you need for the topics you have studied.

You will use your knowledge in two types of case study questions:

- questions that **require** a named example
- questions where your answer will be improved if you give details of a **relevant example**.

Revision flash cards

It is a good idea to make some revision flash cards for your case studies because then you get all the key information you need all in one place.

On pages 113–116 you will find some templates for flash cards to get you started.

Keep these things in mind when you use your case study knowledge:

- Make sure the details you use are relevant to the question. If you write down everything you can remember about your case study, you may waste a lot of time on something that won't count towards your answer

- Named example questions require the name and location of your example. There will sometimes be a special space to write these into your answer but it is best to get into the habit of always naming your example and saying where it is located

- Practise using your case study knowledge in different ways so you can really make it work for you when the right questions come along. Working through past papers is the best way to do this.

Top tips

1 Be careful with scale. If a question wants a named location, then a country or a continent is too large an area to use.

2 Don't confuse less developed countries (poor parts of the world) with more developed countries (rich parts of the world).

3 Make sure your examples are packed with details such as facts, statistics and precise locations!

4 Make sure you use examples that are specific to your case study – don't make general observations.

5 Use geographical terminology that really explains what you mean, for example, 'fertile soil' is OK but 'deep, well-drained soil' is much better.

6 Often there are 3 extra marks available for spelling, punctuation and grammar. Make sure that you take time to read back through your answer to check that it is clear and well organised.

Now try this

List all the case studies you need to revise for your topics. Use the templates on pages 113–116 to help you revise the key details for each of your case studies.

Case study revision cards

How to use these pages

 Find the template for the topic you are revising.

 Identify your case study on that topic.

 Fill in the details on the card and cut it out.

There's only one case study card per topic here, but you need to revise **more** case studies than just these. Use the templates to make your own extra cards.

Once you've made your cards, keep them near your revision notes. Whenever you get bored with your other revision, see if you can learn a case study!

And remember, you only need to learn the case studies **for the topics you have studied**.

You can use the cards at the bottom of p.116 for any topic.

The Restless Earth

Tsunami

Name and date:

Places affected:

Cause

Strength of earthquake:

Names of plates:

Effects

Secondary effects of earthquake

Places close to the epicentre:

Places further away:

Maximum distance affected (km):

People's responses

Immediate

Long term (months or years)

Rocks, Resources and Scenery

How people use .... landscapes

Locations in UK:

Economic uses

Advantages:

Water supply

Advantages / disadvantages:

Farming

Types of farming:

Opportunities:

Limitations:

Tourism opportunities (natural attractions):

Benefits:

Costs:

Challenge of weather and climate

Tropical storm in a .... part of the world

Date:

Name and location:

Causes

Effects

Economic:

Social:

Environmental:

Similarities / differences from rich example:

Responses:

Short term:

Long term (monitoring, prediction, protection, preparation):

Case study revision cards

Living World

Tropical rainforest

Location:

Causes of deforestation

Impacts of deforestation
Economic:

Social:

Political:

Environmental:

Methods of sustainable management

Water on the Land

Dam / reservoir in the UK

Location:

Advantages:

Water transfers / consumption:

Issues connected with the dam / reservoir
Economic:

Social:

Environmental:

Sustainable water supply?:

Overall – are benefits greater than costs?

Ice on the Land

Alpine area for winter sports / glacier sightseeing

Name of area:

Location:

Attractions for tourists

Impacts of tourists
Economic:

Social:

Environmental:

Management strategies

Level of success:

The Coastal Zone

Coastal flooding

Name of area:

Location:

Causes
Why the risk of coastal flooding is high:

Impacts
Economic:

Social:

Political:

Environmental:

Case study revision cards

Population Change
Non-birth control population policy

Name of country:

Strategies used

Positive results

Drawbacks and problems

Overall effectiveness

Changing Urban Environments
Squatter settlement redevelopment

Name and location:

Characteristics

Changes and improvements made

What made them possible and results

Changing Rural Environments
Commercial farming area in the UK

Name of area and location within UK:

Favourable factors for farming
Physical:

Human:

Details about the development of agribusiness

Farming impacts on the environment

Development of organic farming

Changes due to market demands and government policies

The Development Gap
A natural hazard

Name and location:

Causes

Impacts

Responses

Assessment: why poor countries are affected worse than rich countries

Case study revision cards

Globalisation

One type of renewable energy

Type:

How it works; types of location needed

Named examples of locations

Advantages

Disadvantages

Government policies

Tourism

UK National Park or coastal resort

Name:

Reasons for growth / attractions

Management strategies for visitors

Effectiveness of strategies

Plans for the future

Topic:

Name:

Location:

Topic:

Name:

Location:

Answers

UNIT 1
The Restless Earth

1. Unstable crust

1 Pacific Plate, North American Plate, South American Plate, Eurasian Plate, Indo-Australian Plate, African Plate and Antarctic Plate. There are also many smaller plates that you could include in your answer.

2 The North American Plate and the Eurasian Plate.

2. Plate margins

1 The San Andreas Fault.

2 Two from: earthquakes are triggered when built-up pressure is released from inside the Earth's crust by plate movement; at destructive plate margins, energy builds up in the subduction zone and is released as earthquakes; at conservative plate margins, if the plates get stuck, pressure builds up and is released as earthquakes when the plates move again.

3. Fold mountains and ocean trenches

A syncline is a downwards fold in sedimentary rocks, whereas an anticline is an upwards fold in sedimentary rocks.

4. Fold mountains

Two from: HEP requires a rapid flow of water to power turbines to generate electricity; the steep relief of young fold mountain regions means rapid flows of water can be created in HEP plants; high mountains also attract a lot of rain, which drains quickly into the streams and rivers, so there is a regular supply and powerful discharge of water for HEP.

5. Volcanoes

1 Surtsey island is an example of a shield volcano. Etna is an example of a composite cone volcano.

2 A supervolcano is much bigger than a volcano and emits at least 1000 km² of material. They are caused by an enormous magma chamber beneath the crust that pushes the land above it into a dome. After an eruption the dome collapses to create a huge depression called a caldera.

6. Volcanoes as hazards

Primary effects are the immediate impact of a natural hazard. Secondary effects happen later, in the medium and longer term, as knock-on effects of the damage done by the natural hazard.

7. Earthquakes

Answers might:
- focus on the similarity between the distribution of earthquakes and the location of the margins of the Earth's tectonic plates
- note that earthquakes are most frequent along the major destructive plate margins, for example where the Nazca Plate meets the South American Plate
- point out that the map shows most areas away from plate margins as not having earthquakes.

8. Earthquake hazards

This answer will depend on your case studies.
- Give named examples for both earthquakes (state where and when they occurred).
- Focus on **responses** and compare by showing that you understand the similarities and differences in responses. For example, stating what magnitude both earthquakes were would set your answer up for a clear comparison.
- Divide your answer into immediate and long-term responses.
- Give precise details and focus on comparisons rather than descriptions of what happened.

9. Tsunamis

This will depend on the case study you have used here. Knowing the details for all your case studies is a very important part of your revision.

Rocks, Resources and Scenery

10. Rock groups

Layers of sediments are deposited in a sea bed, and as each new layer is added its weight compresses the layers below it until they turn into rock.

11. The rock cycle

Water fills a crack or joint in the rock.

Water freezes and the crack is widened.

Repeated freeze–thaw action increases the size of the crack until the block of rock breaks off.

Loose blocks of rock are called scree.

One mark would be available for the drawings and three for the labels / annotations. Make sure labels are clear and arrow points to the correct stage. Annotation should be sentences which offer explanation.

12. Granite landscapes

The granite weathers slowly so only thin soils are farmed, which are acidic and not very fertile. Because granite areas are often areas of high relief, conditions are not favourable for most kinds of farming: cold and often very wet.

13. Carboniferous limestone landscapes

Dry valleys are old river valleys that formed either when the limestone underneath them was completely saturated with water or when the ground was frozen in the last ice age. Under these conditions, water did not permeate through the limestone but instead could run off the surface as rivers and erode a river valley. Once the water table subsided or the ice in the ground (called permafrost) thawed, the river disappeared underground again as the limestone bedrock became permeable, leaving the valley on the surface dry.

14. Chalk and clay landscapes
Settlements are often found along the boundary between chalk escarpments and the clay vale because chalk was always a more suitable landscape to live on in terms of building materials and grazing land but it didn't have much surface water. The boundary between the escarpment and the underlying clay is where water that has percolated down through the chalk hits the impermeable clay and comes out as a stream: a water source for settlements.

15. Quarrying
This will depend on your case study. You should annotate / label your map to show the **environmental impacts** of the quarry: visibility of location, associated works which could be causing air pollution, changes to drainage patterns, land use around the quarry, roads to the quarry and associated works built through local settlements which could cause noise pollution and damage to buildings from vibrations from heavy trucks. Make sure you can draw a sketch map of your quarry.

16. Quarrying management strategies
Outline means write down the most important points. This will depend on your case study. Include strategies that are being used during the extraction phase (e.g. limited noise from blasting to set hours, planting trees as screens, etc.) as well as those that are used when extraction is completed at the quarry or sections of it. Make clear that these aren't the same as the strategies used during production.

Challenge of Weather and Climate

17. The UK climate
Insolation is the energy from the Sun that is used to heat up the Earth's surface. It heats up lower latitudes more than higher ones because the angle of the Sun's rays is focused on a smaller area at a shorter distance from the Sun at the equator, and the angle of the Sun's rays is focused on a broader area at a greater distance as latitude increases.

18. Depressions
low, rain, warm

19. Anticyclones
A blocking anticyclone is an area of very high pressure, above 1030 millibars, which is strong enough to block frontal depressions from moving across the UK. The anticyclone can hold its position over the UK for several days or weeks and in the summer this means day after day of hot, dry weather. The land heats up, adding to the high temperatures from the summer sun, leading to a heat–wave. Lack of cloud cover means the UK gets the full benefit of the Sun's rays.

20. Extreme weather in the UK
You could make two points and develop them to answer this question, for example: 'Although there is not a proven link between an increased frequency of extreme weather events and climate change, climate change simulations and models do show extreme weather becoming more common as global temperatures rise. And there is also the fact that increases in the frequency of extreme weather events are clearly happening at the same time as global temperature rises.'

21. Global warming
Solar power generation.

22. Consequences of global climate change
Your answer could include the following impacts and consequences:
* water shortages due to a decrease in rainfall, leading to conflicts over available water – especially connected with rights to extract water from rivers for irrigation and HEP
* migrations of people away from areas affected by desertification – leads to conflict because people move into areas other people are already living in (which may themselves be under pressure from climate change)
* conflict over control of resources made newly accessible by climate change, for example oil reserves previously hidden under polar ice sheets.

23. Responses to the climate change threat
The Kyoto Protocol gave the 37 industrialised countries until 2012 to reduce their carbon emissions by an average of 5.2% below their 1990 levels.

24. Causes of tropical revolving storms
Tropical revolving storms are a major threat to human life, property and communication networks. They can cause flooding and water shortages. People may become ill or die from water-borne diseases. Many people may be made homeless when buildings are destroyed by high winds or floods. Cars, roads and railways may be destroyed making communication very difficult and delaying or preventing aid from reaching storm-hit communities.

25. Comparing tropical revolving storms
This will depend on the case studies you have learned for this topic.

Living World

26. What is an ecosystem?
This is the process by which nutrients are released from dead organisms by decomposers and returned to the soil, and then taken up again by producers, eaten by consumers, who then die – and the process repeats itself in a cycle.

27. Ecosystems
Tropical rainforest ecosystems are found north and south of the equator in Central and South America, parts of Africa and down the wet eastern side of Asia.

28. Temperate deciduous woodlands

Sustainable management means using strategies to help the woodland ecosystem meet the needs of people today in a way that doesn't stop people in the future being able to use them in the same way.

29. Deforestation

Biodiversity is a measure of the variety of life forms in a particular area of an ecosystem. Rainforests are easily the most diverse ecosystem on Earth. Any reduction in rainforest area will lead to a reduction in biodiversity and is very likely to mean unique species are made extinct.

30. Sustainable management of tropical rainforest

Tropical rainforests tend to be located in poorer countries and logging, commercial farming and mineral extraction from cleared forest areas all represent good ways for poorer countries to earn money that they can use for development. Because many poorer countries have large international debts, strategies have been suggested and implemented by which poorer countries agree to conserve areas of their tropical rainforests in return for a reduction in their debt payments. These are known as conservation swaps.

31. Rainforest management

Because there is less vegetation cover to shelter the soil from the impact of the heavy rainfall, the soil, which only has a thin layer of organic material, quickly loses its nutrients to leaching down through the soil and through soil erosion as surface run-off increases. If Brazilian ranchers do not take steps to protect their land from these processes, it can quickly become degraded and useless for pasture.

32. Economic opportunities in hot deserts 1

One from: hot desert areas in the US, for example, are popular with retired people primarily because of the hot, dry climate; developers have created special retirement villages which offer security, care and plenty of activities popular with retired people, such as golf.

33. Economic opportunities in hot deserts 2

Water on the Land

34. Changes in the river valley

The long profile shows how the gradient of the river changes as you move downstream. The cross profile of the river shows you a cross-section of the river channel and the river valley.

35. Erosion and transportation

Three from: if the river has to carry extra load it can lose energy; likewise if the gradient becomes less; if the river loses volume (less water in the channel); if the river hits an obstacle or is slowed down by a bend in the river's course. Rivers also slow down at their mouth as they enter a much larger body of water that is not flowing in the same direction; or there are ocean currents where a river enters the sea.

36. Waterfalls and gorges

1   In the upper course of a river, where vertical erosion is dominant.
2   Gorges are formed when a waterfall retreats upstream, cutting a gorge through the band of hard, resistant rock that forms the cap of the waterfall. The resistant rocks keep the sides of the gorge steep rather than sloping.

37. Erosion and deposition

Erosion is involved because as the river meanders in its lower course, lateral erosion erodes away the valley sides, making them wider and removing any interlocking spurs. Meanders migrate down the valley too as the river moves down its course. Deposition is involved every time it floods, when fine particles of silt are deposited onto the floodplain by the flood waters and build up over many thousands of years. Deposition on the inside of meander bends helps fill in these areas as the meanders migrate down slope.

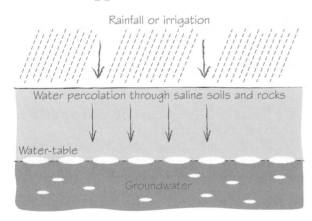

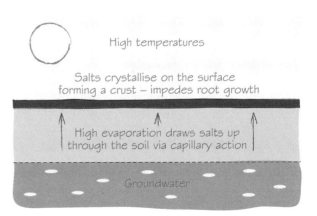

Make sure your diagram is well labelled.

38. Flooding 1

This table shows flood events from 1864 to the present.

The question asks you to list events from 1998.

Date	Flood event and location	Most severe impact
1998	Easter floods, Midlands	5 people killed
2000	Flooding across England	10 000 homes and businesses flooded
2002	Glasgow floods	200 people evacuated
2004	Boscastle flood, Cornwall	100 people airlifted to safety
2005	Floods in Carlisle, Cumbria	3 people killed
2007	UK floods – many areas affected	6 people killed
2009	Cumbria floods	1 person killed
2010	Flooding in Cornwall	100 people evacuated
2012	UK floods – many areas affected	9 people killed Wettest Summer since 1912

39. Flooding 2

1 This answer will depend on the case studies you have done in school.
2 Flooding carries human and animal waste with it, which often contains dangerous bacteria that spread disease.

40. Hard and soft engineering

If an area has lots of river meanders, the water in the river takes longer to move through the area and travels more slowly. A straight channel drains the area more quickly which makes flooding less likely, reduces the amount of water involved if a flood does happen and speeds up the time that the flood waters will take to drain away.

41. Managing water supply

The main problem with using pipelines to transfer water long distances is cost, especially if the water needs to be pumped up a gradient or pipelines need tunnels, etc., to cross high land.

The main problem with river transfers is that the water being transferred usually needs to be from the same drainage basin as the rivers it is being transferred in, otherwise the different chemical composition of the soil and rocks may mean the water being transferred is too different from the usual river water and organisms cannot tolerate it. For both, there is also a political issue – the areas of surplus failing to benefit from the deal while the areas of deficit benefit from a concentration of economic activity. Also some would say the problem with large-scale water transfer is that it does not address problems of leakage and wasting of water in deficit areas, which, if fixed, could make their use of water more sustainable.

Ice on the Land

42. Changes in ice cover

Two from: evidence from landforms and geology which show the areas affected by glaciations and those that were not affected; evidence from fossils, which show where there was vegetation and animal life during glacial periods; also records of pollen deposits which have been preserved in ancient bogs and swamps showing where areas were not covered by ice and where the evidence suggests there wasn't any vegetation because it was all buried under kilometres of ice.

43. The glacial budget

The answer here will depend on what case study you have chosen and the causes for retreat and evidence of retreat that you have studied.

44. Glacial weathering, erosion, transportation and deposition

Smoothing of the bedrock happens when the ice next to the bedrock carries small rock particles. These small particles act like sandpaper on the bedrock, grinding away irregularities to produce a smoother surface. When the ice near the bedrock carries sharper, larger rock fragments, though, these scratch along the bedrock as the ice moves, causing striations.

45. Glacial erosion landforms 1

A tarn / lake.

46. Glacial erosion landforms 2

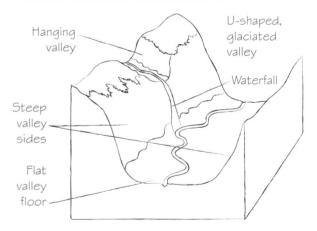

47. Glacial landforms of transportation and deposition

These are end moraines left at the snout of the glacier as it retreats up the valley. There may be several of these: the terminal moraine is the end moraine that marks the farthest extent the glacier reached.

48. Alpine tourism – attractions and impacts

This will depend on the case study you have done for this topic. The revision card for this case study on page 114 will help you with this task.

49. Management of tourism and the impact of glacial retreat

If glaciers retreat rapidly, meltwater will increase very quickly and it might cause flooding instead of providing a reliable water source. HEP systems can usually deal with sudden increases in river flow, but if the glacier were to disappear completely, the river and stream systems that replaced it might not provide enough water for HEP, or the water they provide might not be in the right place for existing systems. The amount of water available for HEP might also be affected by glacier retreat.

The Coastal Zone

50. Waves and coastal erosion

Biological weathering occurs when plant roots widen cracks between the rocks, or animals or birds dig burrows. Chemical weathering happens when rainwater dissolves parts of the rock, for example carbonates in limestone rocks. Mechanical weathering includes processes like freeze–thaw, where repeated freezing and thawing widens cracks in rocks as water expands into ice. Freeze–thaw is most effective on coasts where temperatures regularly dip below freezing point at night.

51. Coastal transportation and deposition

1 Traction, saltation, suspension, solution.
2 It is both. It is formed by deposition but it is shaped by erosion, and if the supply of material to a beach was cut off the beach would erode away.

52. Landforms of coastal erosion

Your diagram should look like the diagram on page 52.

53. Landforms resulting from deposition

Sources such as: sediment carried down to the coast by rivers; erosion of cliffs brought to the beach by longshore drift; material brought up from the seabed by constructive waves. Some beaches are made from material brought in by humans either to replace eroded material or to replace a shingle beach with a sand beach.

54. Rising sea levels

Make sure you can list the economic, social, political and environmental consequences of impacts of coastal flooding in your case study for this topic.

55. Coastal management

Hard engineering
Advantages: methods such as sea walls or rock armour are very effective at stopping the sea from eroding the coastline and may also provide nice places to walk along (sea walls) and good habitats for some fish species (rock armour). Groynes are not as expensive to build and are something we expect to see on the beach rather than an obtrusive presence. They create a larger beach, which is useful for tourist resorts.
Disadvantages: expense – sea walls in particular are very expensive to build and maintain and rock armour can be expensive to transport from where the rocks come from. They are also obtrusive to look at, spoiling the natural beauty of the coastline. Groynes starve beaches further down the coast (in the direction of the prevailing

wind) of sediment so that eroded material is not replaced.

56. Cliff collapse

This example would depend on the case study you have chosen but one example would be the Holderness coast which is retreating at an average of 1.5 m per year. Make sure you revise these points: why some areas are threatened and the impact on people's lives and the environment.

57. Managing the coast

This answer will depend on your coastal management case study. Make sure you have not just described the strategies instead of explaining them, as the question has asked for. And your strategies should be specific to your case study area, not just general strategies that could have come from anywhere.

58. Coastal habitats

Issues will be specific to your case study but they might include: trampling of delicate flowers and vegetation by visitors; damage to creeks by people mooring boats in them or driving powerboats too fast and too close to the habitat; the risk of fires from people dropping cigarettes or lighting fires for barbecues; the need to protect rare bird species from animals like rats and cats which might enter the habitat from nearby housing areas, or from dogs that might disturb nesting birds, or from birdwatchers keen to get a picture of a rare bird; the need to provide parking space for visitors and provide facilities such as cafés and toilets.

UNIT 2 Population Change

59. Population explosion

The number of years from birth that a person is expected to live.

60. Demographic transition

It is called late expanding to show it is a development from the early expanding phase (stage 2). The population is still expanding because birth rates are higher than death rates, but the birth rate drops dramatically through this stage as different factors reduce the number of children women have.

61. Population structure

A broad base.

62. Growing pains

When women have a chance to go to high school and university, they tend to have babies later in life and may also be more likely to know how to use birth control more effectively, if their culture encourages it. More educated women may have greater opportunities in the job market, so women can get better jobs, which has the same result – women start to put off having babies while they are getting established in their jobs and maybe getting promoted. Increased education for women means there is less female illiteracy, and this is likely to lead to changes in a culture that sees women only having value as a housewife and mother to a large family.

63. Managing population growth

Your answer will depend on your case studies. The differences will probably be to do with a government focus on making people have fewer children compared to a focus on moving people from overpopulated to underpopulated areas, or improving the factors that encourage people to have fewer children such as education, access to contraception, female emancipation and economic prosperity/urbanisation. You may also be comparing state control of a population compared to a state looking to encourage voluntary change in its population.

64. Ageing population

Your answer will depend on your case study. The EU sees one solution to the ageing population problem. They suggest allowing more people from outside the EU to come and work and live in EU countries, and allowing people in poorer EU countries to come and work in richer ones. Not only does this boost the working population, but recent immigrants also boost the birth rate too, as has been seen in the UK in the results of the 2011 census.

65. Migration: push and pull

Economic migrants are people who move from one area to another looking for work. It's usually people moving from a poorer country to a richer country.

66. EU movements

Examples might include:
- The Iraq War
- The Afghan War
- In the 1990s, the Bosnian conflict.

Citizens who oppose their government may suffer from political persecution.

Changing Urban Environments

67. Urbanisation goes global

Answers might include:
- disadvantages of urban areas – congestion, high house prices, perceptions of high crime rates
- advantages of rural areas – less congestion, less vehicle pollution, spread of high-speed internet connections to the countryside, cheaper housing, perceptions of lower crime rates.

68. Inner city issues

This answer will depend on what you have studied. The City Challenge strategy in Hulme, Manchester, is a good example. In this case, a partnership between the government and private investors raised £37.5 million, which was spent demolishing unpleasant old housing (some was kept and renovated) and building new housing in pleasant and energy-efficient surroundings; a new park and new schools were also built. The City Challenge strategy was one of the first to make sure the views of local people were taken into account.

69. Housing issues and solutions

Greenfield sites are open land that has never been built on before, so these sites are almost always on the rural–urban fringe. Brownfield sites are areas that have been built on before and are now available for being built on again. Brownfield sites are located within the urban area.

70. Inner city challenges

The CBD is often an 'empty heart' at night because land prices are too expensive for developers to build housing there – the developments are all retail and office spaces because they make most money. However, that means that the CBD is where lots and lots of people work, so all those people have to either commute in from the suburbs to get to their jobs, or find housing in the inner city which may not meet their needs.

71. Squatter settlements

The informal sector is unofficial; people create jobs for themselves and do not pay the government any taxes on what they earn. It does include illegal things (such as selling stolen goods) but often it is just people making things to sell to scratch a living.

72. Squatter settlement redevelopment

Answers will depend on the case study: it would be very good to include a range of development approaches and schemes, including self-help schemes, site and service schemes and other forms of local government help.

73. Rapid urbanisation and the environment

It is difficult because the city is growing much faster than the city authority is able to plan for, and all the people coming in to the city are creating waste. There is unlikely to be enough money to build adequate waste disposal facilities such as landfill sites or incinerators. There are problems with roads: it is difficult for trucks to get access to many parts of the city to collect waste from homes and businesses if the road system has not been expanded to meet growing demand. City authorities may feel they have bigger priorities for the funds available, such as fighting crime, extending education or improving housing. Also, many people in the informal economy depend on other people's waste to make an income.

74. Sustainable cities

Make sure your answer considers a range of sustainable living features, such as:
- Important to conserving the historical environment
- Protection of the environment
- Use of brownfield sites
- Reducing and safely disposing of waste
- Providing adequate open space
- Including people in the decision-making process
- Provision of an efficient public transport system.

Changing Rural Environments

75. The rural–urban fringe

Green belts are areas where planning restrictions protect the rural areas around urban centres from being developed. The idea is that because developers cannot build on the green belt land they will redevelop brownfield sites within the urban area.

76. Rural depopulation and decline

Three common reasons: because second home owners

are not often resident in the village, they are not often there to shop at local shops or use other local services; second home owners do not educate their children in village schools – they go to school where the main residence is; second home owners do not regularly use some local services such as bin collection so these may decline, and they don't often use bus services because when they do come to visit the second home, they drive in their own car.

77. Supporting rural areas

Farm diversification is when farmers use some of their land to do something different from farming, for example: turning a field into a campsite; converting a stable block to a business conference facility; changing the farm into a farm park for visitors to learn about farming life. It helps the rural economy because it can generate more income for farmers, can create more jobs and can increase the range of different jobs available in rural areas, which all helps support rural life.

78. Commercial farming 1

Very big farms mean economies of scale: that is, the agri-business can buy from suppliers in very large volume and get discounts for this that make running the farm cheaper overall. Very big farms can also have very big fields that can be farmed with very big, super-efficient machinery. This maximises productivity.

79. Commercial farming 2

You should consider two advantages and two disadvantages of organic farming. For example:
Advantages: using crop rotations rather than artificial fertilisers; giving the soil time to recover and avoiding the danger of killing beneficial insects as well as pests; organic livestock farming does not use antibiotics and feed additives to keep animals healthy, so these chemicals are not passed on into the environment.
Disadvantages: organic farming does sometimes use some toxic elements, such as copper, in its pest control and these toxins may be just as bad for the environment or worse than chemical pesticides; the organic diet that cattle are fed means they produce double the methane of cattle fed with non-organic feed (since methane is a very powerful greenhouse gas, the environmental impacts of this could be very significant).

80. Changing rural areas: tropical 1

Advantages: (economic) there is a lot more money to be made from becoming a cash crop farmer or selling your land to a cash cropping agri-business; even if farmers cannot benefit from being a cash crop farmer themselves, they can get a job as a worker on a cash crop farm; in the case of Kenya, flower farming is now also a very important way for the country to now earn foreign currency.
Disadvantages: subsistence farmers may lose their land; food production for Kenyan people goes down as farmers turn to cash crops (so Kenya may have to spend more money importing food from abroad); there may be damage to the soil from over-farming; workers on cash crop farms do not earn very much and the work is also seasonal, so they may not be able to get a job all year; farmers who sell their land to an agri-business are left without a means of support; farmers who turn to

cash cropping make more money but also get into debt and this makes life more stressful; cash crop methods may use a lot of water which is then not available to traditional farmers; cash cropping may overuse pesticides or fertiliser, causing environmental problems. Being forced off the land to make way for cash crop farming increases rural–urban migration. Cash cropping to export food to global markets increases CO_2 emissions from transportation and refrigeration.

81. Changing rural areas: tropical 2

Salinisation is caused when high temperatures draw water and salts up through the soil to form a hard crust on the surface. If irrigation does not have proper drainage to wash water down through the soil to wash out the salts then it can make the problem worse. The soil becomes waterlogged and the water table rises closer to the surface, which increases the amount of water and salts travelling up to the surface.

The Development Gap

82. Measuring development 1

You would expect a correlation between HDI and GNI per head, because as income increases per head so should the things that HDI measures, e.g. life expectancy, literacy, education, etc.

83. Measuring development 2

Quality of life is a social measure of development. Instead of just looking at people's income it considers things like access to health care, literacy rates, housing quality, access to safe water, life expectancy.

84. What causes inequalities?

Aim to make two developed points for this answer. Examples include: **prices** – prices for primary products go up and down very rapidly as they are traded in the developed countries, and often they can drop sharply just as a country has spent its resources building up production; **competition** – there is often strong competition between countries producing the same primary product which drives prices down; **processing** – the most profit is made in processing raw materials into manufactured products because they are value-added. The countries that do this are very keen to stop any other countries from joining in; the processing countries also work hard to keep raw material prices down.

85. The impact of a natural hazard

Make sure you refer to a named natural hazard in your answer. The details of your answer will be specific to your case study, but your explanation may include things like: lack of money to prepare defences against natural hazards; lack of money to organise relief effort after disaster; likelihood that many homes and other buildings are not strongly constructed and so may be more likely to be destroyed in the disaster; likelihood that infrastructure such as phone networks, roads, railways and electricity grid are not as developed and therefore more vulnerable to disruption; likelihood that, in urban environments, population density may be very high so localised disasters may affect many more people.

86. Is trade fair?

Debt abolition is when all or part of a country's international debts are written off. Conservation swaps (also called 'debt-for-nature swaps') are when all or a part of a country's international debt is considered repaid in return for the debtor nation investing in environmental conservation.

87. Aid and development

Bilateral aid is aid given directly from one country to another (from one government to another). Multilateral aid is when a government gives aid to an international agency (such as the World Bank), which then puts the aid towards one of its development projects.

88. Development and the EU

Two from: pays farmers a subsidy which gives farmers in poor countries more money; it guarantees minimum levels of production so it boosts agriculture across EU regions; it gives farmers in poorer areas access to the whole EU market and to the world market via the EU. It also removes the 'middle men'.

Globalisation

89. Going global

Possible advantages include: wages are much cheaper in India than in the UK (for example, a salary of £12 000 in the UK compared to the equivalent of £1200 in India); operating costs are also cheaper - the cost of the phone calls themselves and of running a telesales operation in India.

Link these to technological developments: internet making international phone calls very cheap for companies without losing too much phone call quality; the presence in India of around 2 million graduates who can speak fluent English, so call centre customers can still get a reasonable level of service from the India-based call centres.

90. TNCs

There are a range of disadvantages. Two from: profits from TNCs may 'leak' out of the country back to the richer part of the world where the TNC has its headquarters; jobs created by TNCs are not secure because if the TNC finds somewhere cheaper it will move its operation there; TNCs are interested in profit only and may exploit workers if local employment laws do not stop this from happening or bad publicity about exploitation affects sales in richer countries; TNCs often pay low wages and offer jobs in poorer parts of the world that do not develop skills; TNCs will often not be concerned about environmental damage caused by their operations unless local laws prevent it or there is bad publicity about it in the richer countries.

91. Manufacturing changes

Points to develop and link include:
- NIC governments setting up areas where conditions are favourable for new industries, e.g. tax-free zones
- HICs often have a minimum wage which is the lowest wage employees can be paid; NICs often have no minimum wage or very low minimum wage level, so employees can be paid very little
- HICs often have limits on how long working hours can be (e.g. 48 hours a week in the EU); NICs sometimes have no limits on working hours, so workers have to work for long shifts or take on a lot of overtime. This means work is completed faster.
- HICs often have a long history of trade unions, who have helped workers to improve their conditions and can organise strikes to stop production completely if necessary to meet workers' needs; NICs do not often have strong unions and strikes may even be banned – very popular with TNCs
- HICs often have well developed health and safety laws which protect workers from industrial accidents; NICs do not always have much health and safety legislation which makes them a lot cheaper to operate in.

92. China

Explanations might include: China's rapid industrialisation which has also increased China's demand for oil to power its factories and to be used in the manufacture of its products; China's economic development which has meant millions more Chinese each year can afford to buy a car or can afford to travel further by car, increasing consumption of oil; also it highlights China's lack of its own oil resources.

93. More energy!

Increased demand for products that significantly improve quality of life (for example, air conditioning, refrigeration and, most of all, motorbike and car ownership) requires increased energy. Try to show the connection between increasing wealth and an increase in purchases of 'consumer' items (e.g. the latest model of a mobile phone, a new car to replace an older model), the production of which all consume a lot of energy and resources.

94. Sustainable energy use

Your answer should include the central point that if people reduced the amount they bought (e.g. not getting a new mobile phone every time a new model was released), reused what they had more and recycled as much of their household waste as they could, the amount being sent to landfill would be reduced and, more importantly, less energy would be used making and transporting products that most people could really do without.

95. Food: we all want more

Soil fertility is dramatically reduced by overcultivation, overgrazing and lack of traditional agricultural practices such as leaving land fallow to recover.

Tourism

96. The tourism explosion

Two from: culture, history, entertainment, nightlife, stag / hen night venues, romantic locations or other similar options.

97. Tourism in the UK

Two from: positive media coverage of the London Olympic and Paralympic Games around the world; promotions from UK tourist boards that make strong links between London and the successes of the Games; people who visited the Games were impressed by London and wanted to see more of the UK; people who wanted to see the Olympic Park and the sporting venues after the end of the Games.

98. UK tourism: coastal resort

One from: building indoor entertainment facilities that are not weather dependent and can extend the season beyond the summer; upgrading and building better class accommodation to attract a new range of customers from the more luxury end of the market; building conference facilities that will bring custom in all the year round.

99. UK tourism: National Park

Large numbers of tourists can erode footpaths, which can damage the natural habitat and also makes the area less attractive to other tourists. Tourists can also leave litter, which pollutes the area and can be dangerous for wildlife.

100. Mass tourism: good or bad?

The two positive impacts and the two negative impacts you pick will depend on your chosen tropical case study area. For example, for Kenya, you might select the fact that tourists have to pay to get into the National Parks and this money is then used to help run the parks and look after the environment and animals. And because tourists want to see wild animals on safari, it means that most local people understand that the wild animals need to be protected or the tourists will stop coming. For negative impacts in Kenya, there is the damage done to coral reefs off Mombasa by tourist boats dropping their anchors onto the reefs, and the erosion caused to the savannah grasslands by the minibuses that transport tourists around the National Parks.

101. Keeping tourism successful

Your answer will depend on the case study you have done for this topic. The information in this book is about Kenya, but you might have learned some different information for Kenya or you might have done a completely different tropical tourism area case study (e.g. Jamaica), so make sure you revise what you have done in school. 'Outline' means that you should write down the most important points, but remember that this is an 8-mark question.

102. Extreme tourism

Your answer will depend on the extreme environment you have studied. For the Antarctic tourists: sightseeing (from their cruise ships), visit close to the shore in dinghies and small boats, land on the shore (in small numbers) and visit specially designated areas, visit scientific research stations.

103. Ecotourism

Your answer will depend on the example you have studied. Many textbooks give examples of ecolodges in Brazil or Ecuador, or you may have studied ecotourism in the Maasai lands of Kenya (or somewhere else entirely).

Exam Skills

104. Stimulus materials – an introduction

Yellow highlighting for sections the student has lifted straight from the extract, blue for developed statements using their own knowledge.

105. Using and interpreting photos

Meteorologists can measure the size of the storm using satellite images and they can also track the route of the storm by comparing satellite images taken over a period of time. Some satellites are also able to measure the energy given out by the storm.

106. Labelling and annotating

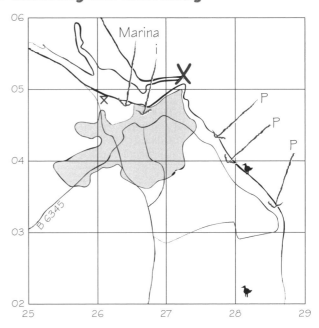

107. Graph and diagram skills

a) line graph; b) scatter graph; c) pie chart

108. Map types

The EU is the largest importer and the USA is the largest exporter.

109. Describing maps

The map of climate zones and the map of distribution of vegetation types.

110. Comparing maps

The same – 1 km (distances on maps do not change with scale, they are just represented in different ways).

111. Exam skills

The correct sentence is: SPaG stands for spelling, punctuation and grammar. And there are only four mistakes, not five.

112. Exam skills: using case studies

Your list will depend on the case studies you have done for each topic.

Published by Pearson Education Limited, Edinburgh Gate, Harlow, Essex, CM20 2JE.

www.pearsonschoolsandfecolleges.co.uk

Text and original illustrations © Pearson Education Limited 2013
Edited, produced and typeset by Wearset Ltd, Boldon, Tyne and Wear
Illustrations by Wearset Ltd, Boldon, Tyne and Wear
Cover illustration by Miriam Sturdee

The right of Rob Bircher to be identified as author of this work has been asserted by him in accordance with the Copyright, Designs and Patents Act 1988.

First published 2013

17 16 15 14
10 9 8 7 6 5 4 3 2

British Library Cataloguing in Publication Data
A catalogue record for this book is available from the British Library

ISBN 978 1 447 94085 2

Printed in Slovakia by Neografia

Acknowledgements
We are grateful to the following for permission to reproduce copyright material:

Logos
Logo of Campaign to Protect Rural England
Maps
Maps of Snowdon/Yr Wyddf; Amble, and Bolton Abbey. Reproduced by permission of Ordnance Survey on behalf of HMSO, © Crown Copyright 2013. All rights reserved. Ordnance Survey Licence number 100030901, and supplied by courtesy of Maps International.

The publisher would like to thank the following for their kind permission to reproduce their photographs:

(Key: b-bottom; c-centre; l-left; r-right; t-top)

A. P. S. (UK): 57; **Alamy Images**: Adam Burton 27c, Aerial Archives 73, AgStock Images, Inc. 32tl, Alex Segre 78l, All Canada Photos 42l, allan ridsdale 77l, Andrew Butterton 74, Andrew Patterson 93c, Ashley Cooper Pics 47, Bernd Mellmann 33cr, Bill Bachman 30t, Catherine Hoggins 55br, Chris Howes / Wild Places Photography 70b, Cosmo Condina Western Europe 42r, darek 69c, David Cole 26c, David Gowans 106tr, David Lyons 41, David R. Frazier Photolibrary, Inc. 23cr, 70t, David Tipling 58, Derek Croucher 16br, Eye Ubiquitous 33tl, FLPA 16cr, 26t, 94r, Frans Lanting Studio 80, Graham Taylor 26cr, Greenshoots Communications 16cl, incamerastock 26cl, 69l, James Davis Photography 98l, Jenny Matthews 81, Jo Katanigra 46, Jon Arnold Images Ltd 45r, Julie Quarry 32cr, Kevin Britland 53br, Kip Evans 27t, lynn hilton 71, Malcolm Park English Coastline 53tr, Mark Humphrey 22, Marmaduke St. John 93r, Michael Willis 16bl, Mike Kipling Photography 36, National Geographic Image Collection 45l, Nick Cockman 15, Pat Tuson 69r, Paul Dronsfield 78r, Robert Stainforth 20, stock_wales 77r, Thom Gourley / Flatbread Images, LLC 32cl, Tips Images / Tips Italia Srl a socio unico 33tr, Tomas Kaspar 27b, travelib prime 53tl, Trinity Mirror / MirrorPix 98r, yu liang wong 93l; **Bahamas Tourist Office**: 96l;
DK Images: Helena Smith 99, Greg Rodin 30b, Jamie Marshall Collection 23bl, Linda Whitwam 13, Patrick Mulrey 23br, Tim Draper 12;
Fotolia.com: Cloudia Newland 105tr, dzimin 105cr, miket 105tl; **Getty Images**: Stu Fraser 97; **Imagemore Co**., Ltd: 94cr;
Les Bell Photography: 106br; **Pearson Education Ltd**: Jules Selmes 66; **Photos.com**: Hanna Liivaar 100; **Science Photo Library Ltd**: NASA 105cl, NOAA 109, Peter J. Raymond 105b; **Shutterstock.com**: AtlasPix 55tr, Georga Green 55tl, Jeremy Reddington 96r, Roha 96c; **Sozaijiten**: 89b; **Veer/Corbis**: aletermi 95, Cepn 89t, pedrosala 94l, 94cl

All other images © Pearson Education Limited

Every effort has been made to contact copyright holders of material reproduced in this book. Any omissions will be rectified in subsequent printings if notice is given to the publishers.

In the writing of this book, no AQA examiners authored sections relevant to examination papers for which they have responsibility.